Advances in Intelligent Systems and Computing

Volume 609

Series editor

Janusz Kacprzyk, Polish Academy of Sciences, Warsaw, Poland
e-mail: kacprzyk@ibspan.waw.pl

About this Series

The series "Advances in Intelligent Systems and Computing" contains publications on theory, applications, and design methods of Intelligent Systems and Intelligent Computing. Virtually all disciplines such as engineering, natural sciences, computer and information science, ICT, economics, business, e-commerce, environment, healthcare, life science are covered. The list of topics spans all the areas of modern intelligent systems and computing.

The publications within "Advances in Intelligent Systems and Computing" are primarily textbooks and proceedings of important conferences, symposia and congresses. They cover significant recent developments in the field, both of a foundational and applicable character. An important characteristic feature of the series is the short publication time and world-wide distribution. This permits a rapid and broad dissemination of research results.

More information about this series at http://www.springer.com/series/11156

Amic G. Ho
Editor

Advances in Communication of Design

Proceedings of the AHFE 2017 International Conference on Human Factors in Communication of Design, July 17–21, 2017, The Westin Bonaventure Hotel, Los Angeles, California, USA

Springer

Editor
Amic G. Ho
Jubilee College, School of Arts
and Social Sciences
The Open University of Hong Kong
Kowloon
Hong Kong

ISSN 2194-5357 ISSN 2194-5365 (electronic)
Advances in Intelligent Systems and Computing
ISBN 978-3-319-60476-3 ISBN 978-3-319-60477-0 (eBook)
DOI 10.1007/978-3-319-60477-0

Library of Congress Control Number: 2017943017

Printed on acid-free paper

This Springer imprint is published by Springer Nature
The registered company is Springer International Publishing AG
The registered company address is: Gewerbestrasse 11, 6330 Cham, Switzerland

Advances in Human Factors and Ergonomics 2017

AHFE 2017 Series Editors

Tareq Z. Ahram, Florida, USA
Waldemar Karwowski, Florida, USA

8th International Conference on Applied Human Factors and Ergonomics and the Affiliated Conferences

Proceedings of the AHFE 2017 International Conference on Human Factors in Communication of Design, July 17–21, 2017, The Westin Bonaventure Hotel, Los Angeles, California, USA

Advances in Affective and Pleasurable Design	*WonJoon Chung and Cliff (Sungsoo) Shin*
Advances in Neuroergonomics and Cognitive Engineering	*Carryl Baldwin*
Advances in Design for Inclusion	*Giuseppe Di Bucchianico and Pete Kercher*
Advances in Ergonomics in Design	*Francisco Rebelo and Marcelo Soares*
Advances in Human Error, Reliability, Resilience, and Performance	*Ronald L. Boring*
Advances in Human Factors and Ergonomics in Healthcare and Medical Devices	*Vincent G. Duffy and Nancy Lightner*
Advances in Human Factors in Simulation and Modeling	*Daniel N. Cassenti*
Advances in Human Factors and System Interactions	*Isabel L. Nunes*
Advances in Human Factors in Cybersecurity	*Denise Nicholson*
Advances in Human Factors, Business Management and Leadership	*Jussi Kantola, Tibor Barath and Salman Nazir*
Advances in Human Factors in Robots and Unmanned Systems	*Jessie Chen*
Advances in Human Factors in Training, Education, and Learning Sciences	*Terence Andre*
Advances in Human Aspects of Transportation	*Neville A. Stanton*

(continued)

(continued)

Advances in Human Factors, Software, and Systems Engineering	*Tareq Z. Ahram and Waldemar Karwowski*
Advances in Human Factors in Energy: Oil, Gas, Nuclear and Electric Power Industries	*Paul Fechtelkotter and Michael Legatt*
Advances in Human Factors, Sustainable Urban Planning and Infrastructure	*Jerzy Charytonowicz*
Advances in the Human Side of Service Engineering	*Louis E. Freund and Wojciech Cellary*
Advances in Physical Ergonomics and Human Factors	*Ravindra Goonetilleke and Waldemar Karwowski*
Advances in Human Factors in Sports, Injury Prevention and Outdoor Recreation	*Tareq Z. Ahram*
Advances in Safety Management and Human Factors	*Pedro Arezes*
Advances in Social & Occupational Ergonomics	*Richard Goossens*
Advances in Ergonomics of Manufacturing: Managing the Enterprise of the Future	*Stefan Trzcielinski*
Advances in Usability and User Experience	*Tareq Ahram and Christianne Falcão*
Advances in Human Factors in Wearable Technologies and Game Design	*Tareq Ahram and Christianne Falcão*
Advances in Communication of Design	*Amic G. Ho*
Advances in Cross-Cultural Decision Making	*Mark Hoffman*

Preface

Communication of design is concerned with understanding users, creating communication, and engaging experiences. Users expect communication to be an exciting and interactive experience. Therefore, to provide such enjoyable and stimulating experiences, researchers in the discipline of communication design have identified that human factors try to enrich communication and optimize design experience. Topics covered in this book report on humanistic approaches, including the roles of human needs, emotions, thoughts, and actions, and have revealed more innovative approaches based on visuals. Furthermore, these types of connections influenced users' experiences in design consumption. Such experiences generally exerted a considerable effect on users' satisfaction toward the designs. Considering human factors in the communication of design enables designers to be actively connected with human needs.

The exploration of human factors and design in the past decades is an appropriate and valuable opportunity to enrich and strengthen the field of communication design. This book is organized into two sections that focus on the following subject matters: Communication of Design Methodologies and Communication of Design Applications.

This book will be of special value to a large variety of professionals, researchers, and students in the broad field of communication of design and human–computer interaction, who are interested in feedback of devices' interfaces, user-centered design, and design for special populations, particularly the elderly.

Each section contains research paper that has been reviewed by members of the International Editorial Board. Our sincere thanks and appreciation to the board members as listed below:

We hope this book is informative, but even more that it is thought provoking. We hope it inspires, leading the reader to contemplate other questions, applications, and potential solutions in creating good designs for all.

July 2017 Amic G. Ho

Contents

Communication of Design: Methodologies

Humanistic Teaching Pedagogy in Typographic Communication Design

Amic G. Ho(✉)

School of Arts and Social Sciences, The Open University of Hong Kong,
Ho Man Tin, Kowloon, Hong Kong, China
amicgh@gmail.com

Abstract. General training of typographic communication design focused on skill-based practices. Few scholars recognized the importance of conducting continuous academic research in theoretical and methodological aspects and on how to bring the humanistic elements and intellects to typographic design education. In this study, the research team found that ethnographic research method enabled designers to establish new aesthetic perspective and approaches that supports the fusion of cultural and functional qualities for contemporary typographic design. Eventually, the new teaching method increased the awareness to cultural sustainability and the possibility of experimental design. Methods adopted in this study aimed at exploring the effective approach for learning typographic design in undergraduate education. Participants' feedback and reflection were collected and reflects that ethnographic research was an effective learning experience for elaborating the undergraduate students' knowledge and skills in typographic design from the living environment to humanistic typographic communication design.

Keywords: Emotion · Teaching pedagogy · Ethnographic research · Typographic communication design

1 Introduction

Generally, the teaching methods of typographic communication design focused on skill-based training. However, students were found that only obtained the capability of using computer fonts literally typing out tag lines and slogan for their design work. Such finding indeed shocked the typographic communication design for the fact that student's insufficient knowledge and trust to the power of typographic communication design and typeface design. It was essential to equip the visual communication designer to have sufficient knowledge and skills in typographic design. The structure of typeface, especially for Chinese, was complicated, no matter it is hand-generated or computer-generated. There was a need to include a comprehensive learning on the principles of using the typefaces and creating new typefaces. However, few educators focus on the education of typographic design and few studies focus on optimize the learning process of it. This study aimed to examine if ethnographic research would be an effective tool for enhancing design students which provide more supportive sources for their typographic concept development. It expected that investigated result of this

A.G. Ho (ed.), *Advances in Communication of Design*, Advances in Intelligent Systems and Computing 609, DOI 10.1007/978-3-319-60477-0_1

study in typographic design would be generalized to the cases of other languages those are heavily related to the daily living of humans.

2 The Current Problem of the Typographic Design Education

There are some problems occurred in the current typographic design education. The study conducted in 2012 examined the understanding level of design students on Typographic design [1]. They found that less than 30% of the students got experience on creating new typefaces, and around 30% presented that they never got any experience on creating typeface for their work. The scholars compared the feedback obtained from the participants in 2011 and 2006 [1], they found that the more student got experience on creating new typefaces in 2011. These finding show that the importance of typographic design education was recognized in the ethnographic of design education. Some educators started to introduce the concept of typography and the methods of creating new typeface. However, the increasing awareness on typography could not catch up with the actual needs in the ethnographic of industry. According to the investigation which interviewed some design consumers, around 70% of them agreed that that they needed to have unique and eye-catching typographic design for their product and promotional items [1]. The underlay reason was the arising market in China. The market needed creative and experimental typographic design those provided surprises visual experience to the consumers [2]. Less than 25% consumers think learning typographic design were unnecessary and resources wasting. At the same time, the nowadays design students relied much on the internet sources. The internet sources undoubtedly provided large amount of information. However, there was no guarantee on its reliability. Some of them often contain mistakes or wrong concepts. Typographic design teaching and learning emphasize continuous investigation and production of new meaning through resource organization [3, 4]. Similar to the other design disciplines, students who learn typographic design are encouraged to equip themselves with the capability to collaborate and generate solutions to complex design problems [5]. Typographic design education not only encourages students to produce design outcomes but also cultivates the characteristic spirit of the culture in which the design students live, thereby facilitating the development of their thinking as designers [6, 7]. Furthermore, the traditional approaches to literacy education, a recognized shift towards student-centred learning that encourage students to take part in design communities actively.

3 The Importance of Research in Typographic Design

Many typographical studies focus on design processes. Few designers focus on conducting initial design requirements before they start their projects. Initial requirements study is considered one of the most important methods to discover the potential for typographic design considerations [8]. As there were too many elements in contents designed the digital world, the expressive potential of typeface design have to be explored by the designers. Scholars proposed that the research process was a preparation period to investigate the potential executions of the typographic design [8]. This was an important step to let the creators of the messages widening the variety on typographic execution to enhance their communication. However, designers had been ignored the role of designer research on typographic ethnographic. There were many examples across communication, social science, and design ethnographic proved the contribution of research process to the optimization of the typographic design. Some other scholars proposed the similar concept [3]. They pointed out that living nurture typographic design. Research activities not only work for identifying audience and the channel of communication. The research process in typographic design leads designers to understand the limitation of the existing limitation of the communication. Designers had to explore the solutions to break these limitations down. This exploration brought the new interruption of languages and words. They provide opportunities to enhance the creativity. Designers learn from their investigation in life and experience. They should be trained to enrich their own horizontal of observation and experience. Their point of view was similar to encourage the manipulation of ethnographic research in typographic design education [3]. This recognition on the contribution of typographic was different with the general bias on typographic design, those only considered typographic design in limited contexts, for example, selecting typefaces for the commercial design [9], executing for the branding campaign and graphic design for commercial".

4 The Manipulation of Ethnographic Research in Current Education

The effectiveness of adopting ethnographic research as a research method is highly recognized by many scholars [10]. It provides original sources for revealing inedited findings. It is one of the most important instruments for research work, for example, Ph. D. thesis, journal and books [11]. However, ethnographic research is not commonly adopted in the design studies, especially on the undergraduate level. The original sources of ethnographic research reveal founded facts those solely resulted from ethnographic research. In the process of ethnographic research, achieved documents, historical documents and diplomatic inscriptions, concluded interviews, questionnaires, records from focus groups based on the research domain. Generally, the ethnographic research process in education include: the formulation of research focus, study design, negotiation on potential research sites, documents analysis, observation and interview in the ethnographic research [10, 11]. Key advantages of ethnographic research are its accuracy and updated investigations and it have important influence on the research outcomes [12]. The manipulation of ethnographic research in education was mainly for encouraging design students to experience and ask questions. Based on the investigations of students' experience in education, scholars proposed six steps of the science learning process: recognizing situations, clarifying problems, formulating hypothesis, testing hypothesis, revising with further tests and acting on solutions [13]. The National Research Council supported this point of view, therefore, it questioned the teaching quality provided by the science teachers in the general science curriculum based in school. It emphasized that science should involve "…thinking like a scientist. The science process was developed as the guideline of learning included 'observing, classifying, inferring, controlling variables'…. [14]" Scholars interpreted some learning skills (i.e., individual process skills) as a part of teaching strategy [15]. The influence of conducting ethnographic research in teaching and learning experiences on graduate students. Their findings suggested that a cooperative interaction between simultaneous engagement and research skill development can be generated through conducting ethnographic research in teaching and learning at the graduate level. Graduate students describe, explain, and support their research processes and outcomes. Through these learning experiences, students tend to show increased improvement in scientific reasoning and writing skills [16]. In addition, teachers or instructors involved in research tend to be more effective in engaging students by using ethnographic research instruction. Moreover, the connection between research and teaching was strengthened when graduate students teach in areas directly related to their research ethnographic. Graduate students develop their self-concept as a teacher and researcher [15]. Furthermore, graduate students' understanding of their strengths and weaknesses in research competency can be enhanced through teachers' feedback. Consequently, the students' overall achievement in the graduate program was enhanced. The ethnographic research approach in teaching and learning provides students with a strong foundation for research at the postgraduate level and lifelong learning, which was the "ongoing, voluntary, and self-motivated" pursuit of personal experience or professional knowledge [17]. The learning process of doctoral training in different subjects. They demonstrated the importance of research in doctoral training, revealing that doctoral training entails preparation for scholarship, in which the developing scholar both conducts and reports the research [18, 19]. Scholars observed the learning process of students in her study through her visual arts education web, which based on the ethnographic research in teaching approach. She found that students worked collaboratively with teachers and enjoyed their learning process when the teacher carried out questions to stimulate students' thinking. Scholars proposed to conduct ethnographic research in learning process from the mathematics and science education perspective [20]. He identified the following skills and processes that must be developed: 'knowing and understanding; estimating', 'computing, and solving'; 'visualizing and modeling', 'representing and communicating'; 'conjecturing', 'reasoning', 'proving', and 'decision-making'; and 'applying and connecting'.

5 Scenario Analysis Method on the Manipulation of Ethnographic Research in the Typographic Design Education

5.1 Research Aims

Scenario analysis on the manipulation of ethnographic research in the typographic design education was conducted to understand how the ethnographic research contributed to investigate the relationship of everyday life and typographic design. Thirty design students participated into this ethnographic research in Beijing. All participants were undergraduate students who had not less than 2 years' experience in manipulating design projects. They had sufficient design experience to manipulate their design process independently and at the same time, they were still open to try design and research methods in alternative ways. Ten of them learned and followed the provided guidelines to conduct the ethnographic research and the other ten design students manipulated the design process without any guidelines. In both two groups, two design students were paired up as a design and research group. g new design and research methods.

5.2 Research Process

The ethnographic study was conducted for three weeks (i.e. twenty-one days) and included five main stages: interview, observation, document gathering, tutorial, design process. The Fig. 1 showed the schedule of the stages. From the first to third days, teams were invited to have ethnographic research included interview, observation, document gathering for creating typographic design. The participants understood the influence of historical and social changes on typographic design the research. The participants visited various places in Beijing for understanding the historical and social changes of Beijing city. Tutors shared with the participants about the experience on creating new typefaces in workshop. Tutors provided Team A guidelines of ethnographic research on the investigating the relationships between daily living, personal experience, historical background and typographic design in the tutorial. The participants in Team A practiced the research methods of ethnographic research, which included observation and original sources collecting. Team B was invited to process their design research with their own methods. The participants developed them to a completed concept and they execute the design in the following eleven days (i.e. the stages of design process. After obtaining the inspiration from the ethnographic research, Team A designed sets of new typefaces based on the inspiration obtained from the ethnographic experience. On the twenty-first day, the two groups presented their research findings, their typeface design concepts and design outcomes.

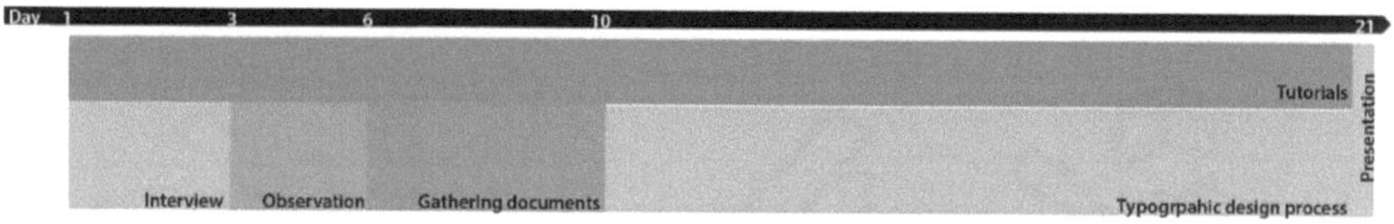

Fig. 1. Scheduled learning activities.

5.3 Scenario Analysis Samples

One group of the students from Hong Kong teams visited the old residential area in Beijing (Fig. 4). Their attention was caught by the decommissioned military factory buildings those boasted the unique architectural style. She got inspiration from the style, its arrangement of the space and forms. She tried to learn from the structure of the old buildings and applied them as the structure principle of her new typeface design (Figs. 2 and 3). Their observation in the ethnographic research provide her the inspiration and the photo, as original sources, took during the observation provided her principles for constructing the typefaces design.

Another scenario analysis sample 2 from Team A, was developed by the inspiration obtained from the public area of under constructing residential areas in Beijing (Fig. 7). They explored the public space of the buildings and interested in the spacing usages and construction parts of the buildings, such as the forms of walls, and stairs (Figs. 5 and 6). They obtained primary sources by photo taking and analyzed them.

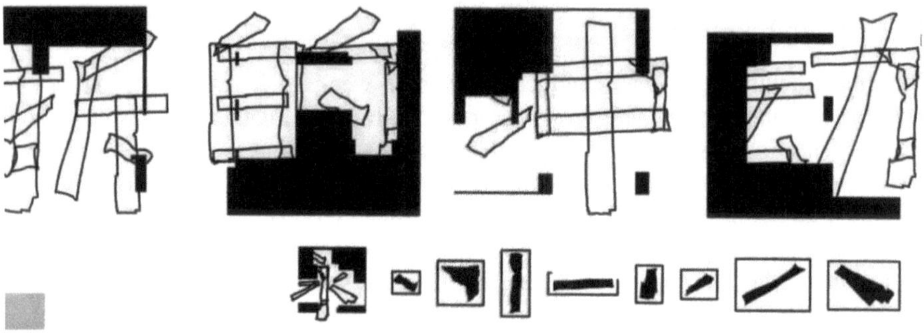

Fig. 2. Typeface design of Team A which based on the ethnographic research in the old residential area in Beijing. (Sketches drew by students Helena Chou, during Beijing old residential area observation, launch on 10 May 2016 at Beijing)

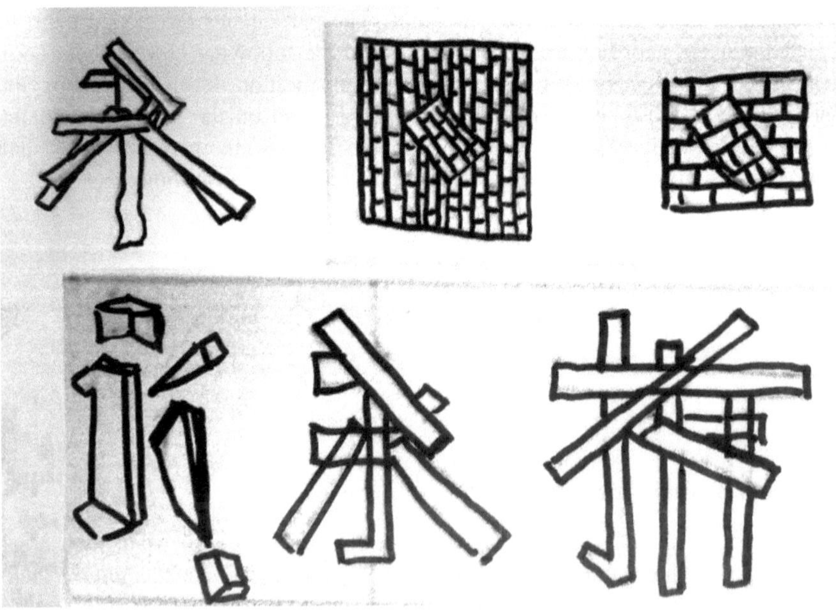

Fig. 3. Sketches of typeface design of Team A which based on the ethnographic research in the old residential area in Beijing. (Sketches drew by students Helena Chou, during old residential area observation, launch on 10 May 2016 at Beijing)

Fig. 4. Typeface design of Team A which based on the ethnographic research in the old residential area in Beijing. (Photo took by students Helena Chou, during Beijing old residential area observation, launch on 2 May 2016 at Beijing)

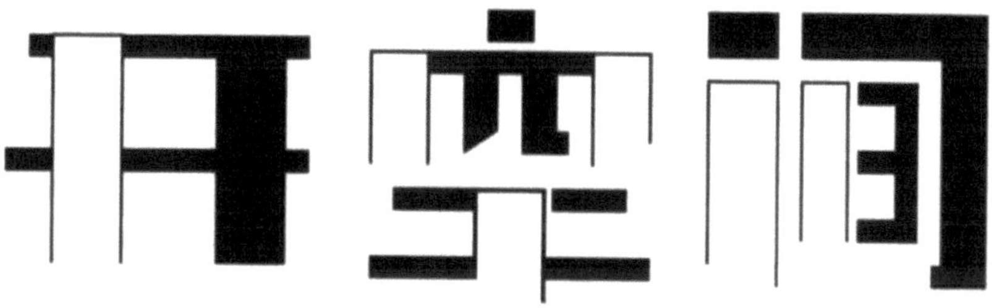

Fig. 5. Typeface design of Team A which based on the ethnographic research in under constructing residential areas in Beijing. (Sketches drew by students Anna Tan, during Beijing observation, launch on 9 May 2016 at Beijing)

5.4 Research Result

Based on the finding from this study, only Team A students provided their design process for scenario. Another group of students form the Team B conducted their research based on the secondary sources included books and resources from the Internet. Compared the two scenarios from Team A and Team B, the design outcomes from Team A obviously contained the investigated historical elements those obtained from the ethnographic research. The investigated elements worked well and enhanced the innovation of the designs. That means participants could explored insights from the original sources in-depth and these insights provided the foundation for the typeface's concept generations. Compare to the design outcomes form Team A, Team B developed design concepts based on the secondary sources. They were easily being influenced by the existed concepts rather than investigating a new one by themselves.

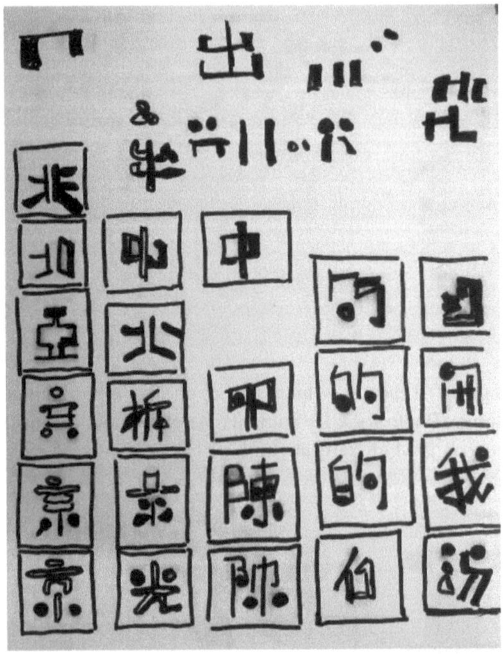

Fig. 6. Sketches of typeface design of Team A which based on the ethnographic research in under constructing residential areas in Beijing. (Sketches drew by students Anna Tan, during Beijing observation, launch on 9 May 2016 at Beijing)

Fig. 7. This photograph of under constructing residential areas was captured on 14 May 2016 by student Anna Tan.

6 Conclusion

To conclude, using scientific research in design education was important in equipping students for their design learning and their further professional works in creative industry. We found that investigations on the primary sources encouraged the learning intention of the students and their research ability in the design curricula. Ethnographic research provided much more inspired insights than the internet research as it leaded students to investigate most accurate, and up-to-date information. It was more useful to design students in their design process which relied on original sources. This is much more workable than internet sources which provide information those provided limited guarantee on the accuracy and documented quality.

References

1. Choi, J., Chiu, M., Sylvia, T.: RE-TRANS-formation of Chinese typography. In: Between black and white – ATypI Hong Kong (2012)
2. Meggs, P.B., Purvis, A.W.: Meggs' History of Graphic Design. Wiley, Hoboken (2016)
3. Liu, E., Lu, K.Y.: The Teaching and Learning of Chinese Fonts Design. Huazhong University of Science and Technology, Wuhan (2010)
4. Kress, G.: Literacy in the New Media Age. Routledge, New York (2003)
5. Liem, A.: Using design education to survive in the 'corporate world' of higher learning and research. J. Des. Res. **9**(2), 104–118 (2011)
6. Games, I.A., Squire, K.: Design thinking in gamestar mechanic: the role of gamer experience on the appropriation of the discourse practices of game designers. In: International Society of the Learning Sciences. Utrecht University, The Netherlands (2008)
7. Gee, J.P.: Situated Language and Learning: A Critique of Traditional Schooling. Routledge, New York (2004)
8. Billard, T.J.: Fonts of potential: areas for typographic research in political communication. Int. J. Commun. **10**, 4570–4592 (2016)
9. Thomas, S.: Designing Obama: A Chronicle of Art & Design from the 2008 Presidential Campaign. Post Press, Chicago (2010)
10. Merriam, S.D.: Qualitative Research and Case Study Applications in Education. Jossey Bass, San Francisco (1998)
11. Edmonson, S., Irby, B.: Ten Tips for Producing a Top Qualitative Research Study. Pearson, Toronto (2008)
12. Serban, I., Illie, C.: The use of ethnographic research in the educational process. In: Proceedings of 5th World Conference on Educational Sciences, vol. 116, pp. 3231–3235 (2014)
13. Dewey, J.: Science as subject-matter and as method. Science **31**, 121–127 (1910)
14. Berman, E.: Sándor, Gizella, Elma: a biographical journey 1. Int. J. Psychoanal. **85**, 489–520 (2004)
15. Maher, C.A., Powell, A.B., Uptegrove, E.: Combinatorics and Reasoning: Representing, Justifying and Building Isomorphisms. Springer, New York (2009)
16. Maher, C.A.: A Study of the Influence of a Preservice Science Teacher Education Program over Time. ProQuest, Ann Arbor (2009)
17. Kamler, B., Thomson, P.: Helping Doctoral Students Write: Pedagogies for Supervision. Routledge, London (2006)

18. Ng, A.H.S.: The creation of the 'Hong Kong Visual Arts Education Web' and the use of the inquiry-based teaching approach. Int. J. Art Des. Educ. **28**(2), 215–224 (2009)
19. Sleeswijk, V.F., Stappers, P.J., van der Lugt, R., Sanders, E.B.N.: Contextmapping: experiences from practice. CoDesign: Int. J. CoCreation Des. Arts **1**(2), 119–149 (2005)
20. Med, M.: DepEd K to 12 resource guide for teacher educators, school administrators and teachers. In: Preferred Education Solutions Provider in Southeast Asia and Beyond (2012)

Directing Untrained Users' Attention Using Simple Sound Patterns

Lars Bo Larsen[1(✉)], Søren Staal Jensen[2], and Mai Baunstrup[3]

[1] Aalborg University, Fredrik Bajers Vej 7, 9220 Aalborg, Denmark
lbl@es.aau.dk
[2] Timeplan, Vandmanden 10C, 9200 Aalborg, Denmark
sjensen1984@gmail.com
[3] Visma, Karenlyst Alle 56, 0277 Oslo, Norway
baunstrup@gmail.com

Abstract. In this work, we show how simple sound patterns can be used to direct a person's attention towards specific locations, when the sound is emitted from a single, fixed position. In interaction design, this is an effect most often achieved using visual ques, such as animations or flashing lights. Using sound instead is useful in situations where visual ques are undesirable or unfeasible. We designed two experiments with a total of n = 45 respondents. 10 different sound patterns were composed with inspiration from previous studies and used as stimuli. The results showed significant differences ($p = 0.05$) for directing the gaze upwards and downwards and trends for left-right. The composed sounds require no prior training, or knowledge of cultural references (contrary to e.g. earcons and auditory icons) by the respondents and can thus be regarded as universal in nature.

Keywords: Human factors · Interaction design · Auditory displays · Sound design · Directing attention

1 Introduction

Human interaction with technology today is highly based on displays and the use of the visual sense. However, there are different social and work situations in which a flashing or dynamically changing display would be inappropriate. This could be either in accordance to social acceptability or in situations where the visual senses are occupied elsewhere, such as automotive contexts. Furthermore, not all devices have a visual display. Consequently, there has also been much research into the field of using sound in interaction with technology, such as sound scape design, virtual 3D sound, use of earcons and auditory icons. In the work presented here we are interested in replacing the visual modality with the auditory in certain situations.

A common problem in interaction design is how to capture and direct the users' attention towards a particular location, e.g. on a screen or device front plate. A simple and well-used solution is to use dynamic visual elements, such as a blinking LED on a device front plate or an animated icon, as known from common OS GUIs.

A.G. Ho (ed.), *Advances in Communication of Design*, Advances in Intelligent Systems and Computing 609, DOI 10.1007/978-3-319-60477-0_2

In the present work, we focus on the ability of sound to solve this problem. The solution corresponding to the visual scenario described above would be to use directed sound, either emitted from a particular physical location or a virtually generated 3D sound, to draw the users' attention towards the sound source. This is further discussed in Sect. 2.

However, this solution imposes some undesirable constraints and requirements on the design-space, both seen from the designer's and the manufacturer's viewpoint. Regardless whether real or virtual localized sound is used, the scenario calls for a number of speakers to be located at strategic positions on or around the device. The user will also be required to be located in a particular position or area relative to the device emitting the sound. This is necessary to correctly perceive the sound source location. This will make it more difficult to achieve a good design, restrict the number of users to one at a given time and will make production of a device more complex and expensive compared to a single-speaker solution.

Discarding both the real and virtual localized sound solutions, we will instead turn towards investigating the design of particular sound patterns. In other words, we will use the sound's intrinsic characteristics instead of the localization of the sound in an attempt to direct the user's spatial attention. This is similar to what is sought by earcons and auditory icons, except here we are looking strictly for the *perceived spatial characteristic*. We expect this will introduce certain constraints regarding e.g. the nature and duration of the designed sounds, but will nullify the need for specific spatial placement and number of speakers, as well as the users' position.

As a brief outline of the paper, the next section describes related work in the field of designed sounds, which can guide the user in certain directions. Following this, an experiment with specific sound patterns will be described. Finally, the results and discussion are presented together with the conclusions drawn from the study.

2 Related Work

One of the first important studies in this field is from 1930. In an experiment, Pratt placed loudspeakers behind a screen at five different height intervals with a fourteen-point scale in front of it [1]. The loudspeakers emitted tones at 256, 512, 1024, 2048 and 4096 Hz. Pratt then played the tones back in randomized order and location. The subjects were told to indicate where they thought the sound came from. The result of the experiment was that high frequency tones were generally perceived higher up in space than the tones of lower frequency, regardless of the physical location the tones were emitted from [1, 2].

In a later experiment Roffler and Butler (1968) placed four loudspeakers on top of each other behind a screen divided into thirteen numbered sections. During the experiment test-subjects sat 1.5 meters from the screen, and they were told to indicate where they thought a sound originated from, via a number on the scale. The sounds used were pure sine tones and varied from 250 to 7200 Hz, with a duration of 10 ms. The results were similar to those found by Pratt [2, 3]. Other researchers have reported similar findings [4, 5].

More recently (2000), Walker and Ehrenstein [6] carried out experiments, in which they explored whether sound with an increase or decrease in pitch, could be perceived as having a direction. Twelve auditory stimuli were produced in the ranges 316 Hz to 2512 Hz. The auditory stimuli all had an onset pitch and then either increased or decreased in pitch to indicate either an "up" or "down" sound pattern. All sounds had durations of 150 ms. The accuracy of the responses was greatest (96%) when the pitch change was congruent, in the sense that the pitch either started high and increased or started low and decreased. This was compared to the 87% accuracy of incongruent pitch change [6].

One of the few examples of successful designed sounds, that guided people left or right was done in a study by Walker in 1987 [7]. 828 subjects were tested to see how they matched acoustic features (frequency, waveform, amplitude, and duration) to visual metaphors on a response sheet. The subjects were asked to listen to sounds designed in accordance to the mentioned acoustic features. The subjects were then told to match these sounds to either: shape, size, pattern, vertical or horizontal (via choices on the response sheet). Sounds which were based on duration were consequently matched with horizontal (i.e. left-right), and frequency with vertical (up-down). The duration sounds were designed by either using a pure or complex tone, with the frequency kept constant [7].

Thus, research into the perceived spatial location of sound patterns has been ongoing for many years. Most of these studies were designed with the presupposition that a certain sound should be associated with a location or even a metaphor presented visually. Thereby, the subjects were primed with a certain level of prior expectation about matching sounds to locations, or directing them to push certain areas to map a location for the sounds. Other studies contain acoustic experiments explaining guidelines, with distances and angles, from where a person can locate the position of the actual sound source [2].

2.1 Use of Sound Patterns in Interaction Design

William Gaver (1986, 1989) discussed how sounds are interpreted by humans and how this can be exploited in interaction design [8, 9]. In particular, he introduced the concept *auditory icons*. Gaver focused on the interpretation of the sound pattern, in contrast to the studies mentioned above, which focused primarily on the psychoacoustic perception of the sound pattern and not the semantics a sound may convey to humans.

Auditory Icons. Gaver defined auditory icons as "*caricatures of naturally occurring sounds such as bumps, scrapes, or even files hitting mailboxes*" [8, p. 169]. The principle is similar to visual icons and relies on the users' ability to map the sound from a physical world to the semantics of the auditory icon. In other words, the use of auditory icons relies on a certain level of cognitive effort by the user. Either as a conscious mapping of unlearned auditory icons or a recall of already learned ones [8, 9].

Earcons. Blattner (1989) and later Brewster (1994) investigated earcons [10, 11]. Earcons are small musical messages. The building-blocks of earcons are motives, which are short melodies used as individual recognizable entities. Using rhythm and different combinations of the same motive it is possible to create a whole family of earcons, which has its own understandable hierarchy. These can be efficient and highly recognisable (such as the MS Windows sign-on jingle or error message). However, there are also certain disadvantages. First of all, earcons are based on abstract musical messages, with no mapping to real-world sounds. This means that the user has to learn and retain the meaning of them. They cannot use their knowledge of the world in order to figure out what a given earcon should represent, unlike auditory icons.

A combination of auditory icons and earcons has also been described [9]. This is achieved by modifying auditory icons according to the principles pertaining earcons.

It is important to note that both auditory icons and earcons rely on a semantic interpretation of the sound content. Initially, when learning the semantics (especially true for earcons) and subsequently, when they are encountered in auditory interfaces. Thus, there is a certain cognitive effort associated with both earcons and auditory icons. Auditory icons might need less learning and recall, but are more context dependent and studies have indicated that earcons are perceived as more pleasant [12].

2.2 Goals of the Present Study

Having briefly outlined how humans spatially perceive certain sound patterns and how the semantics of sounds can be characterized and used in interaction design, we will now describe the goals and motivations behind the present study.

The use of multiple sound sources (loudspeakers) to create real or simulated sound source locations are discarded due to the drawbacks imposed by the positional requirements described above. We instead turn to exploiting the semantic content of the sounds. However, while it has been shown that a spatial message to some degree can be encoded in auditory icons they still impose a certain cognitive load on the user, when decoding the meaning. Furthermore, they require the listener to have a specific prior experience (or a specific enrollment session) and maybe even a certain cultural background to enable decoding.

The aim of this study is therefore to explore whether a sound pattern can guide a person's attention in a certain direction, without any prior contextual knowledge or training and with a single sound source. By this, we aim to exploit low-level cognitive processing and eliminate any learning or recognition requirements. This requirement would exclude single-pitch tones, such as the studies reported in [1, 2] as these are based on comparisons between the tones and thus require prior exposure to the sounds. However, we chose to include single pitch tones in the study for comparison to the other stimuli in the experiment described in Sect. 3.4 below.

We carry out the study through an experiment, where the respondents perceived location of the emitted sound patterns is observed and recorded. Their subjective experiences are obtained by a post-test questionnaire and interview. The experiment is described in Sect. 3 below and the results in Sect. 4. In the following we will denote the sound patterns as D-sounds to indicate the intended directional capabilities.

3 Experimental Approach

We design an experiment to explore, if certain sound patterns (D-sounds) can guide an untrained persons' attention to a certain location or area on a surface, independent of the actual sound source localization. Furthermore, we also want to find out if the context of the D-sounds can override the acoustical directions from where the sounds are actually coming from. The experiment is divided into two parts, A and B as described below.

3.1 Description of the Experimental Setup

The experiment utilises a custom-built device consisting of sixteen loudspeakers arranged in a 4×4 grid. The loudspeakers are hidden from the respondents behind a grey cloth screen. The device is shown in Fig. 1. The respondent is placed in an adjustable chair one meter in front of the device with her head centred at the same height as the centre of the speaker device.

Ten different D-sound patterns (see Sect. 3.4 below) were played at a normal listening level in random order, and the respondent was asked to point at the speaker device with a laser pen at the location, where she perceived the sounds were emitted from. 64 sounds were played in a row. After each series of 64 sounds the respondents

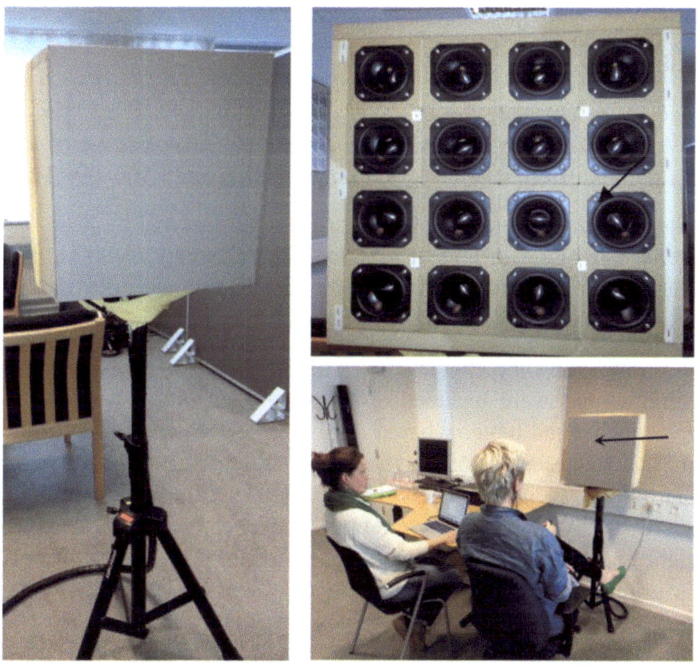

Fig. 1. Left: loudspeaker device used in the experiment. Top right: close-up of the 16 individual speakers without the cover. Bottom right: facilitator with computer and respondent pointing with the laser at the speaker device (laser dot at arrow)

were given a small break to prevent overstimulation and fatigue. A questionnaire was handed out right after the experiment and a short explorative interview was conducted for each respondent.

3.2 Respondents

45 adults participated in the experiment. 28 were male and 17 female, aged from 18 to 61. Everyone reported to have normal hearing and not experiencing listening problems. We did not submit the respondents to an audiometric screening test, as the task did not include a broad range of frequencies and the loudness was deemed sufficient for persons without hearing impairments. The sound level could be adjusted, if a respondent requested it, as this was not an experiment variable, but nobody did.

3.3 Materials

The equipment used in the experiment consisted of: A custom built array of 16 2" loudspeakers arranged in a four by four grid (see Fig. 1), roughly measuring 45 × 45 cm, a custom built multi-channel Bang and Olufsen Ice Power amplifier, a TC Electronics pre-amplifier, controlled via a MATLAB program running on a MacBook Pro. This was also used for data recording and analysis.

3.4 Stimuli

The independent variables are the D-sound pattern stimulus and the speakers emitting the sound. A musician composed ten sound patterns. They consist of simple piano tones with a duration from 0.9–1.7 s and pitch varying from 100–900 Hz. These are:

- Two patterns with increasing or decreasing tempo, inspired by [7].
- Two hybrid patterns, with both tempo and pitch change.
- Two pitch change patterns, inspired on the results reported in [6].
- Four patterns with high, medium high, medium low and low pitch, inspired by [1].

Table 1 below shows the characteristics of the D-Tones

The hybrid sounds were included to investigate whether it would be possible to make a "diagonal" sound pattern, based on the up and down sounds of pitch change and the left and right sounds of duration. Five different loudspeaker configurations were used in order to facilitate the control and the experimental conditions, see Fig. 2 below. Four of the five loudspeaker configurations were used as a control condition: Here the eight loudspeakers in either the upper, lower, left or right half of the 4 × 4 loudspeaker grid were used. For the experimental condition, all sixteen loudspeakers were active at the same time. The 10 D-sounds were played 2 times across each of the 4 control loudspeaker configurations. The D-sounds were played 8 times across experimental configuration. The control condition was included in order to reveal whether the D-sound patterns could override the acoustics in the users' response.

Table 1. The 10 D-sounds used in the experiment.

Name	Duration (Sec)	Tones	Frequency (Hz)	Tempo (Bpm)
Tempo up	1.7	8	444	250–444
Tempo down	1.7	8	444	440 180
Diagonal up	1.7	10	109–882	191–666
Diagonal down	1.7	10	882–129	740–192
Appagio up	1.5	10	139–887	400
Appagio down	1.5	10	887–142	400
High 2	0.9	1	882	–
High 1	0.9	1	444	–
Low 2	0.9	1	221	–
Low 1	0.9	1	107	–

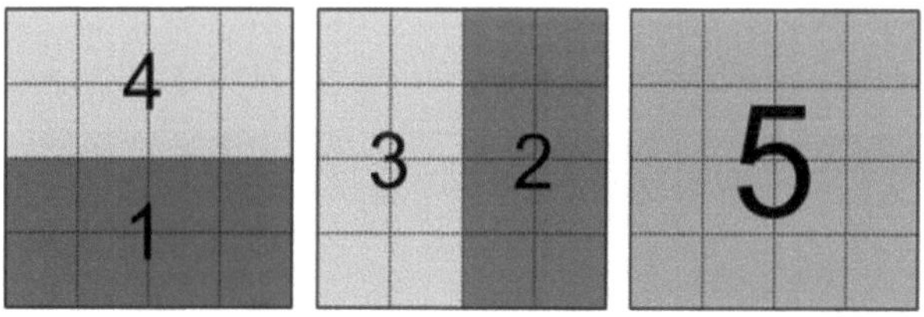

Fig. 2. Areas 1–4 indicate the bottom, right, left and top speaker subsets. Area 5 indicates all speakers are active

3.5 Procedure

The respondents were introduced verbally to the experiment and asked to sign a consent form. The experiment was split into two parts, A and B. 30 respondents participated in part A, where the D-sounds were either rendered by all the speakers, or a subset of them. See Fig. 2 above. The purpose of experiment A to verify whether the respondents were capable of identifying the location from which the sounds were emitted and how this influences the D-sound effects. These were emitted either from a specific area (1–4) on the speaker device using a subset of the loudspeakers, or from all of the loudspeakers simultaneously (5). The stimuli were presented in random order.

15 new respondents participated in part B, where all loudspeakers (5) were active throughout and the D-sounds were emitted uniformly across the surface of the device. Thus, the only spatial clues present are due to the sounds themselves. Every time a sound is played, the respondent would use the laser pointer to indicate where s/he perceived the sounds' location to be. This is observed by the facilitator and entered on the experiment PC by clicking in a 16 × 16 grid shown on the screen. As soon as the entry is recorded, the next sound is played.

4 Results

Overall, part A of the experiment shows that, when a subset of speakers is used (area 1–4), the respondents were indeed able to detect this and identify the locations reliably. However, the individual sounds also played a role. We use heat maps corresponding to the 4 × 4 response grid to depict the results.

Figure 3 shows an example of the responses to the "Appagio Down" or decreasing pitch D-sound. Comparing the responses from areas "1" and "4" shows the combined effects of the sound as well as the position. When the D-sound is only emitted in the upper area (4), respondents still perceive it as mainly coming from the bottom area, but with a tendency towards the middle of the device. The heatmap for area (5) confirms this. Area (2) right, and (3), left, shows the same tendencies, although respondents clearly are able to detect from which side the sound is emitted. This is confirmed in the exit interview; without being asked directly, many of the participants expressed that they quickly learned to distinguish between the spatial locations of the sound emissions to make their response.

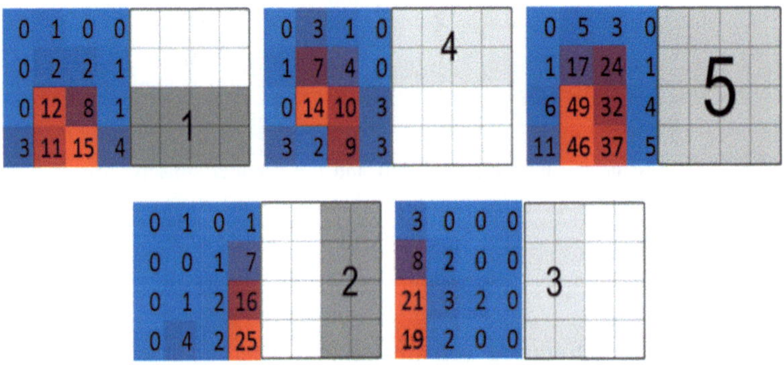

Fig. 3. Heatmaps for "Appagio Down" (Decreasing Pitch) for areas 1–5 for Experiment A.

We saw a similar tendency for the increasing pitch D-sound and the high, mid-high, mid-low and low static pitch D-sounds. However, the D-sounds with tempo changes and the combination D-sounds did not yield any clear results. Thus, even though experiment A confirmed the ability to direct the perceptual experience of the sound source location, it also documented that a habituation, or learning effect had occurred and no clear conclusions were made.

In order to eliminate the impact of the physical location, as well as the observed learning effect, all speakers were active in Experiment B. Thus, the respondents did not get any spatial information to base their responses on. In this case, CHI-square tests did indeed yield significant ($p = 0.05$) results for the decreasing and increasing pitch changes (Appagio Up and Down), and the static high, medium-high and low pitch D-sounds. The D-sounds with increasing or decreasing beats, intended for left-right directional effect, as well as the combined sounds indicated some trends, but did not yield significant results.

85% of the responses for the "Increasing Pitch" D-sound were situated in the upper half of grid and 88% of the "Decreasing Pitch" responses were situated in the lower half. Figure 4 below compares the heat maps along the corresponding Heatmaps from experiment A.

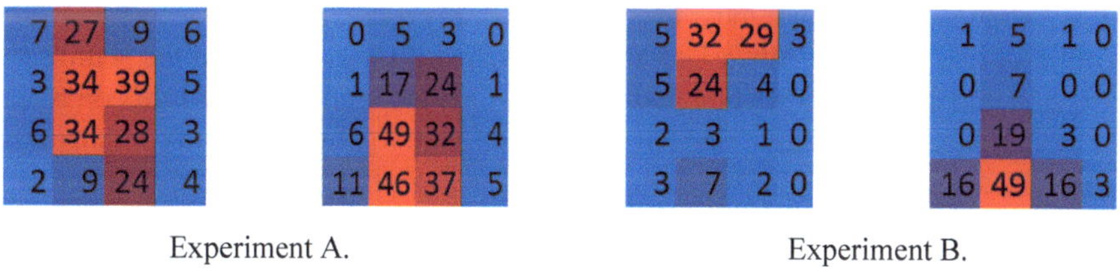

Fig. 4. Heatmaps for increasing (left) and decreasing (right) pitch for experiment A and B for area (5).

When asked, about the right-left effects in the exit interview, half the respondents (53%) confirmed they had experienced the effect, despite non-significant results in the observed data. The corresponding results for the up-down effects were 100% for "up" and 93% for "down".

The impact of the learning effects in Experiment A is evident when comparing the Heatmaps from the two experiments.

5 Discussion

The results from Experiment A indicate an effect of the D-sounds, even when the respondent is aware that the sounds are emitted from different locations on the speaker device. The effect was strongest for up-down direction, whereas right-left direction was less pronounced. However, no significant results were obtained in this experiment. This might be due to a learning effect, where respondents became aware of the different speaker configurations (1–5) and this influenced the results.

Experiment B was carried out with all 16 speakers active at all times. Thus, any observed effects can only be due to the respondents' perception of D-sounds. In this case, significant results were obtained for the up-down direction, but not the right-left direction. We ascribe the learning effects to the different results obtained in Experiment A and B.

5.1 Impact of Chosen Sounds

Clearly, the specific design of the D-sounds must have an impact. In the present case, piano tones were used as they were considered to be neutral in the sense that they are not used as auditory icons or earcons in the most common computer systems. Furthermore, they are considered pleasant to listen to. The individual notes were clearly distinguishable. The duration of the D-sounds was up till 1.7 s. This is fairly high compared to the studies mentioned in Sect. 2. If the sounds were to be used in an interactive system, the duration will most likely need to be reduced to less than half a second.

No clear results were achieved concerning the increasing and decreasing tempo, intended to induce a right or left direction. This effect is obviously harder to achieve with a simple pattern and may even be culturally dependent on e.g. the reading direction of one's culture. We asked all respondents, if they associated the sounds with a direction and about 50% replied that they did. One respondent replied that on a piano the keys are laid out with low tones to the left and high to the right. Thus, an increasing pitch indicated a left-to-right direction to her. However, this was the only comment we received regarding this issue.

5.2 Impact of Observation Method

The location of the laser dot used to indicate the respondents perceived location of the D-sound was recorded by the facilitator observing and noting which of the 16 sub-quadrants on the speaker device the dot appeared. There is some uncertainty associated with this. However, the two facilitators running the experiment trained extensively for this task and pre-trial recordings verified they were reliably able to do this with 98% accuracy.

6 Conclusion

The goal of the present work was to investigate whether it is possible to draw a persons' attention in a certain direction, regardless of the exact position the D-sounds was emitted from. Moreover, the respondents must have had no prior training, not draw on cultural references or previously learned sound patterns or be able to compare with other sounds.

The A and B experiments confirm this is possible to achieve for the up-down directions, using increasing or decreasing pitch patterns. Significant statistical results were obtained and close to 100% of the respondents stated they had experienced the effect for this case. Less clear results were achieved for the right-left direction, although half of the respondents replied they had associated the effect with the D-tones.

We conclude that sound patterns with a clear in- or decrease in pitch can indeed be used to direct an untrained users' attention up- or downwards. This is useful for a broad range of products and situations. In particular, for devices offering a number of spatially distinct "interaction points", such as ATMs, gasoline pumps and modern parking meters. These are often equipped with a small loudspeaker placed within the body for sound feedback. Here D-sounds can help to achieve a better user experience, lower production costs compared to implementing visual elements and most likely a simpler and more aesthetic design.

6.1 Further Work

Additional studies could be carried out to identify D-sounds with a perceived left-right directional effect, as this was not demonstrated fully in our experiment. Eye-tracking could be used to investigate in more detail, whether the responses are reflected or un-reflected. Finally, as this was a lab experiment, it would be useful to validate the results in a real-life situation: By designing D-sounds and include them in a usability trial with naïve users, measures such as task completion times and user satisfaction could reveal, if the desired effect is indeed present.

Acknowledgements. We received financial support for this work from Aalborg University and Innovation Fund Denmark, through the Danish Infinit network. Musician Keld Bauman composed and rendered the D-sounds for us. Bang and Olufsen A/S provided the loudspeakers and amplifiers used in the experiment. Finally, we thank all our respondents for lending us their ears and time.

References

1. Pratt, C.C.: The Spatial Character of High and Low Tones. Journal of Experimental Psychol. **13**, 278–285 (1930)
2. Blauert, B.: Spatial Hearing – The Psychophysics of Human Sound Localization. The MIT Press, Cambridge (1997). Revised edition
3. Roffler, S.K., Butler, R.A.: Factors that influence the localization of sound in the vertical plane. J. Acoust. Soc. Am. **43**(6), 1255–1259 (1968)
4. Mudd, S.A.: Spatial stereotypes of four dimensions of pure tone. J. Exp. Psychol. **66**(4), 347–352 (1963)
5. Simon, J.R., Mewaldt, S.P., Acosta Jr., E., Hu, J.-M.: Processing auditory information of two population stereotypes. J. Appl. Psychol. **61**(3), 354–358 (1976)
6. Walker, B.N., Ehrenstein, A.: Pitch and pitch change interact in auditory displays. J. Exp. Psychol.: Appl. **6**(1), 15–30 (2000)
7. Walker, R.: The effects of culture, environment, age, and musical training on choices of visual metaphors for sound. Percept. Psychophys. **42**(5), 491–502 (1987)
8. Gaver, W.W.: Auditory icons: using sound in computer interfaces. Hum.-Comput. Interact. **2**, 167–177 (1986)
9. Gaver, W.W.: The SonicFinder, an interface that uses auditory icons. Hum. Comput. Interact. **4**, 67–94 (1989)
10. Blattner, M., Sumikawa, D., Greenberg, R.: Earcons and icons: their structure and common design principles. Hum.-Comput. Interaction **4**(1), 11–44 (1989)
11. Brewster, S.A., Wright, P.C., Edwards, A.D.N.: A detailed investigation into the effectiveness of earcons. In: Kramer, G. (ed.) Auditory Display, pp. 471–498. Addison-Wesley, New York (1994)
12. Gaver, W.W.: Auditory Interfaces. Handbook of human-computer interaction (1997)

Looking Forward to Reverse Image Search: Measuring the Effectiveness of Reverse Image Searches in Online Help

Fer O'Neil[1,2(✉)]

[1] Texas Tech University, Lubbock, USA
fer.oneil@ttu.edu
[2] ESET North America, San Diego, USA

Abstract. When performing an internet search for help content, people use one of several possible methods, but user preferences tend toward basic, text-based search queries using common search engines. However, computing and digital device trends show a preponderance of image and multimedia use among most users. This trend reveals a gap in the information design/technical communication discipline for understanding why users are unaware of alternative search methods, and why information design/technical communication is not investigating ways to develop the alternative image search methodology for help content and systems. Specifically, there is a lack of research on the usability and user preferences of image search for online help content. This research will provide an overview of the reverse image search technology, how we can optimize existing and future content to prepare content for its use, and a discussion about why this technology represents the future of user search experience.

Keywords: Human factors · Content-based image retrieval · Usability · Information design with human factors · Media and communication design

1 Introduction

When performing an internet search for help content, people use one of several possible methods, but user preferences tend toward basic, text-based search queries using common search engines. However, computing and digital device trends show a preponderance of image and multimedia use among most users, for example in site-specific rich media searches such as Amazon's "personalized image recall" and Pinterest [1]. This trend reveals a gap in the technical communication discipline for understanding why users are unaware of alternative search methods, and why technical communicators are not investigating ways to develop the alternative image search methodology for our help content and systems.

Specifically, there is a lack of research on the usability and user preferences of image search for help content. As the literature review below will discuss, searching for online help using a reverse image search can be a more effective and satisfactory method for search, but most people do not know about it. This paper proposes a quasi-experiment to help answer the following question:

A.G. Ho (ed.), *Advances in Communication of Design*, Advances in Intelligent Systems and Computing 609, DOI 10.1007/978-3-319-60477-0_3

- When people are made aware of reverse image search as an alternative search method, are their search results as effective as traditional text-based search queries?

To answer this question, the quasi-experiment will include a usability study to test users performing a traditional text search, followed by an intervention that explains and demonstrates a reverse image search, concluded by another search using the reverse image search method. Also included is a post-test survey to measure the perception of satisfaction. The purpose of this research is to investigate how users perform and rate the two search methods and whether the image search intervention will show a significant indication of a more effective search experience.

2 Review of Image Use and Content Searching Trends

The next generation of internet users is increasingly using images and concomitant technologies to share and search for information. Current trends in technical communication include the use of visuals in documentation and their use (such as using screenshots in content) will only continue to grow [2–4]. This trend toward more visual use by people and in documentation presents unanswered questions regarding alternative methods for delivering content to users. An abundant source of research on this topic is available within the technical communication discipline, because the dissemination of information is an integral component to producing technical content [5, 6].

Search is an important method of information discovery and online help content consumption. Existing research, usability studies, and analytics indicate that users typically seek information using their preferred search engine. For example, Yeh et al. researched information seeking on the web and found that users may prefer public search engines to dedicated online help (such as contextual help or using a site search) "because the web may have fresher, more accurate, more abundant information than the built-in help" [7].

Though many users are not familiar with the alternative search methods, usability research of image-based searches indicates that it is just as effective and easy to perform as traditional text-based searches [7]. The next three sections review the literature on this topic and explain the various search methods, their current use, usability features, and how image search technology fits in to current search trends. Following the review of the literature and prior research, the remaining sections will explain this research paper's methods of analysis, participants, results, discussion, and finally a summary and conclusion of the notable aspects of the research.

2.1 Usability of Search

Many disciplines use usability studies (such as A/B testing, user task analysis, surveys, etc.) to test the efficacy of site search. Usability is the combination of effectiveness, efficiency, and satisfaction with which the users accomplish defined goals in a given environment [8].

There are many established guidelines for adhering to and measuring usable features of search. The website Usability.gov provides a basic overview of both site search and search box features. For example, keyword searches should display the same text on the resulting web page that was searched and users should be notified when multiple (i.e., alternative) search capabilities exist [9]. Other research supports these guidelines (in fact, the Usability.gov website and content is derived from existing, secondary research) and also provides additional insight for creating and implementing usable search experiences [9–11]. For instance, research from many disciplines has examined SERP qualities by measuring both the technology behind the search as well as the users performing the search. The connection made between the technology and users is important, and current research in user-centered design addresses how the perception of the usability of a system or technology (measured as effectiveness) correlates to higher usability scores (measured by performance), where users' mental models of how systems behave are examined for their realistic use in actual test cases [12]. Below, this paper will look at the usability of search experiences using content-based image retrieval (CBIR) and examine how it differs from current search trends.

2.2 People Do Not Use Advanced Search

Most people use public search engines as their primary search engine, relying heavily on the main search box [13]. Advanced search options, such as refining based on categories or databases (i.e., faceted search) is a common area of research in information science, as it is most often used in the specialized databases of libraries [14]. The published research on advanced search options shows similar challenges, use, and results to image searches. The most significant takeaway regarding advanced search options is that usability decreases as search features increase, and even when people try to use advanced features, they usually fail because they misuse the technology. The research from Tran et al. and Choi [13, 14] compared different advanced search methods (such as keyword search, faceted search, and query completion), and their results showed that information retrieval was a significant factor primarily because people are not good at using search features.

2.3 Searching by Image is More Effective than Text-Based Searching

The recent research on the topic of CBIR searches for problem-solving (or "know-how" articles) provides the most compelling evidence that general users can perform image searches with the same, or better, results and perceived satisfaction. Two usability studies, one in 2009 and the other in 2011, from Yeh [15], demonstrated that a large, general corpus of images can be efficiently and effectively searched using CBIR techniques. Additionally, the user study showed that searching by screenshot is both easy to learn and faster to specify than keyword searching [7]. The research questions from Yeh's study are relevant to the research question and purpose of this paper's present study. Using a large, general corpora of images in a custom database, Yeh investigated the following two inquiries:

1. Screenshot queries are faster to specify than keyword queries, and
2. Results of screenshot and keyword search have roughly the same relevance as judged by users [7].

The results of the usability test showed that the "average query formulation time was less than half as long for screenshots as for keyword queries" [7]. The later usability study from Yeh [15] expanded the original prototype design into a more robust CBIR system, and introduced an additional text query to help refine the CBIR search.

A similar technology is available using Google's CBIR search when an image fails to return sufficient (i.e., effective) SERP results. The text field "describe image here" is a text query that refines the "IMG_1154.jpg" CBIR query (see Fig. 1). The later (2011) study from Yeh is the most relevant and applicable research to date to focus on the technical (accuracy of CBIR techniques) and usability (perceived system usefulness) aspects of real-world image searches [15]. The study conveyed the efficacy of searching for graphical user interface (GUI) elements, questioning features, or troubleshooting an error, with the contextual help that is traditionally offered in current software applications. These options should offer the best choice, but current help documentation is typically limited in this regard, prompting users to search for help using either public search systems (e.g., Google) or the company's site search.

Fig. 1. Google image search text refinement.

3 Methods

The research conducted for this study was submitted to the Texas Tech IRB and approved as exempt (2015).

3.1 Participants

The participants included the first 10 adults (minimum age is 18) who replied to a scripted email inviting them to participate after they read an informed consent statement. I sent the email to colleagues at my work; therefore, this is a convenience sample. Additionally, the participants must have met the one other demographic criteria to qualify as a participant of not having used reverse image search more than once in the past three months. I sent the email to 10 people, and I received 10 qualifying replies.

One advantage to a within-subjects study, such as this one, is that this type does not require as large a sample size as other types of studies, and there is no need to calculate differences across groups [16] or to focus on the results of individual participants [17]. Additionally, I do not foresee a Hawthorne effect in my experiment because I performed the testing at the same computer, using the same procedure, and nothing was significantly different to have to account for [18].

3.2 Procedures

The 10 selected participants were asked to complete a usability test (Task A and Task B) to measure effectiveness, efficiency, and satisfaction in accordance with ISO 9241-11. The post-test survey used the "SUS-A quick and dirty usability scale" [19], as this is the most widely used test and is considered the standard in usability studies [20, 21].

The single ease question at the bottom of the survey, or SEQ, measured satisfaction and is used in usability research and cited by Nielsen and others [22, 23] along with Sauro and Dumas [24].

There were three parts to the usability testing: Task A, a traditional text search; an intervention to explain the process of reverse image search; Task B, a reverse image search.

4 Results and Discussion

This section includes a brief explanation of the methods of analysis used to measure the results of the usability test. Afterward, the results of the testing and analysis is detailed, including a discussion of the data as it relates to answering this paper's research question.

4.1 Methods of Analysis

To answer the research question for this study, I used the following metrics and associated methods of measurement. Tables 1 and 2 display the methods of analysis used for determining the effectiveness of the two task queries.

Table 1. Performance metrics and methods used

Performance metric	Method used
Time to complete task	Timed task, in seconds
Success rates	Task complete (0 or 1)

Table 2. Perception metrics and methods used

Perception metric	Method used
Ease of use/usability	SUS
Satisfaction	SEQ (Likert)

As stated above, there is no one measure of usability or effectiveness, and this evaluation must be made by combining methods, such as those in this study.

4.2 Summary of Collected Data

Table 3 displays the mean SUS score, mean satisfaction rating, and completion rate. When the task-level data is presented in the next section, I will explain how I calculated the completion rates and what those mean when analyzed with task-time and task-level satisfaction ratings.

Table 3. Summary of mean SUS and satisfaction and both task completion rates

Task	Mean SUS	Mean satisfaction	Completion
Text query	89.8	4.7	75%
Image query	66.3	3.8	91.60%

4.3 Tasks A and B Search Results

On average, Task B (image search) participants completed the search task quicker, as evidenced by a lower mean (25.01 to 30.14) and range (18 to 64.1).

Figure 2 shows the individual task times for Task A, along with a confidence interval. I calculated the interval using 1 standard deviation using a .05 alpha, or 95% confidence level.

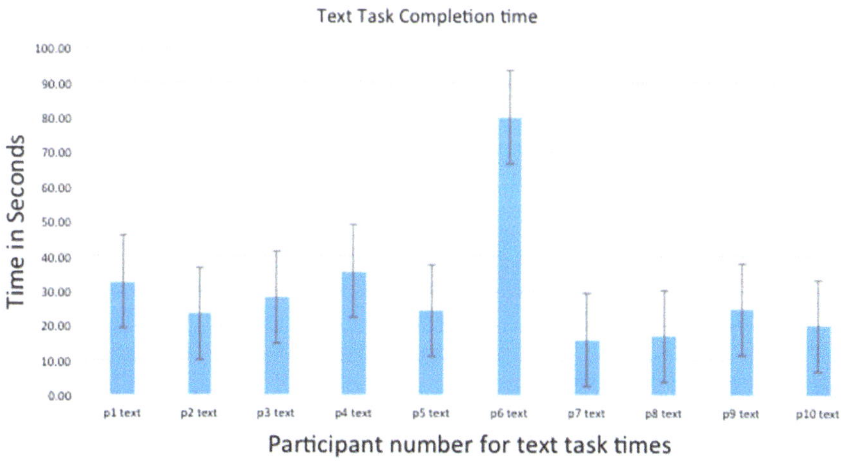

Fig. 2. Text task completion times.

Of note is that the text completion times had much higher variance than the image times (346 and 29.5, respectively) as shown in Fig. 2 above. The image search task also had a closer mean than the text time (25 and 30, respectively) and higher confidence level (which translates to a smaller confidence interval) calculated to 3.89 versus 13.3 for the text confidence level.

The confidence level for the image task times was calculated with the same 1 standard deviation with a .05 alpha, or 95% confidence level. The image search task times and confidence intervals are displayed in Fig. 3 below. Measuring the task times is merely the first step to analyze the tasks for effectiveness, and the following sections use the total task time results to investigate possible correlations with other measured metrics.

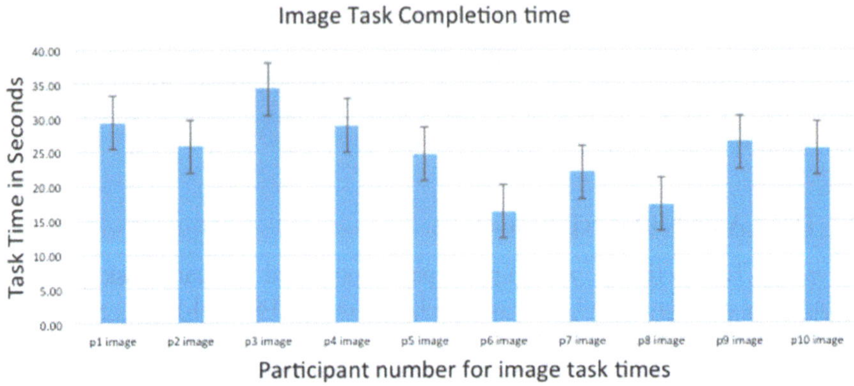

Fig. 3. Image task completion time.

4.4 Correlation Between Task Time and Satisfaction Rating

I analyzed the task times with the satisfaction rating to see if the data allows for any assumptions about the results and their meanings. Figure 4 is a linear regression graph that shows the relationship between the Task A task time and the satisfaction rating. It has a correlation coefficient equal to negative 0.62. However, a best practice for

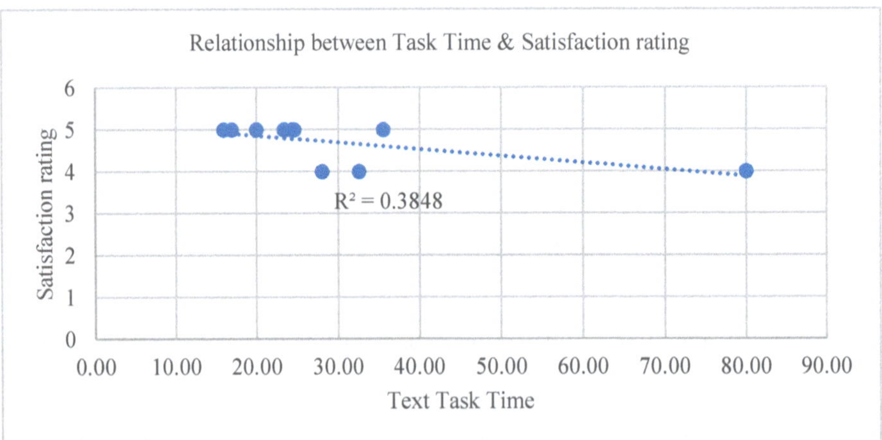

Fig. 4. Task A text search correlation between task time and satisfaction rating.

displaying this kind of data is to calculate the R-squared value, which is a measure of the strength of the relationship between the two variables [16]. The R-squared value "measures the proportion of variability in a data set that is accounted for by a statistical model" [16] and in general, the higher the R-squared, the better the model fits the data. This is important, because knowing one of the two values makes it possible to predict the other one.

The negative correlation coefficient means that as satisfaction rating increases, text task time decreases. The .38 correlation also tells us that there is a strong relationship between the satisfaction rating and text task times.

Figure 5 shows the scatterplot for the relationship between task time for Task B and the participant's satisfaction rating. This has a correlation coefficient equal to negative 0.27 and an R-squared value of .0775.

The negative correlation coefficient is the same as Task A and means that as task time increases, satisfaction rating decreases. Comparing the correlations between Tasks also indicates that there is a stronger relationship between satisfaction rating for Task A (text task times).

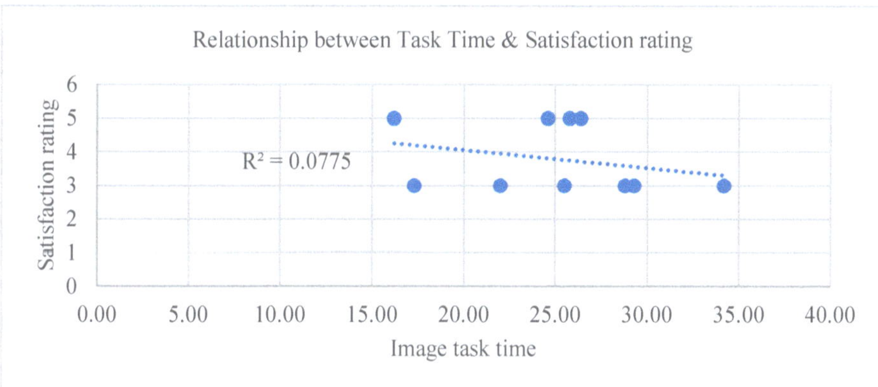

Fig. 5. Task B image search correlation between task time and satisfaction rating.

The low correlation score, coupled with a concomitant low variance with task completion times, seems to reflect the prevailing research conclusion that novel experiences tend to receive lower satisfaction scores, even when task times are better [25]. This research also looks to include completion rates in the analysis of total task times and satisfaction ratings, as the following section will elucidate further.

4.5 Completion Rates

The completion rates for the tasks were calculated because these have been shown to correlate with the other prototypical usability metrics of task-time and task-level satisfaction, as referenced above. The Task A completion results show that two participants did not find the correct, pre-determined search result (.75% adjusted Wald).

Completion rates are the fundamental usability metric—as stated earlier, if users cannot complete a task, other metrics will not provide much insight.

However, around 14% of users who fail a task still rate it as easy to do, and research suggests that some users do not know they failed the task, while others complete the task but still think it is difficult. This perception and lack of awareness may account for an incongruity between higher usability and satisfaction scores and lower completion rates [26]. Task A from my study seems to be an example of this from the completion rates and confidence intervals for each task. I used a formula derived from Sauro and Lewis's 2005 paper [26] and used the formulas recommended in the research on this topic for each task and data type.

The completion rate for Task B, on the other hand, was 100%. Calculating the confidence for a 100% completion rate is a little different from the calculation for Task A and this is because of my sample size. I relied on Lewis and Sauro's article "When 100% really isn't 100%: improving the accuracy of small-sample estimates of completion rates" (2006) to determine the calculation. Because it is unlikely that any general population would ever achieve a 100% rate, a slightly modified formula was used to determine the completion rate–in this instance, it is 91.6%.

Remember, the task completion rates are important because these have been shown to correlate with the other usability metrics of task-time and task-level satisfaction. However, as the data shows, there seems to be an incongruity between usability scores and task completion rates that aligns with the current research on this topic (see Procedures section). That is, the completion rate and the correlation between SUS score and satisfaction rating is higher for Task B, which may be a better indicator for evaluating the effectiveness of the search tasks.

4.6 Correlation Between SUS Score and Satisfaction Rating

The last metric I used to help answer my research question was the relationship between the SUS scores and the satisfaction rating. The Task A results show a correlation coefficient of positive 0.69. This correlation suggests that there is a strong relationship between SUS scores and satisfaction rating. You can see in the Fig. 6 scatterplot that as SUS increases, satisfaction scores increase as well. Though the image results demonstrates this better because there is less variance, I think that a larger sample would show similar results with the Task A, text search test.

The Task B results are shown in Fig. 7. The scatterplot shows the relationship between SUS score and satisfaction rating and it has a correlation coefficient of positive 0.83 and an R-squared value of .69. This correlation suggests that there is a strong relationship between SUS scores and satisfaction ratings, which means that this is a good indicator for determining usability. Finally, the analysis of the correlation of SUS scores and satisfaction ratings seem to indicate that along with task time and task completion, the Task B image search query was more usable, even though the participants gave lower self-reported SUS scores and satisfaction ratings.

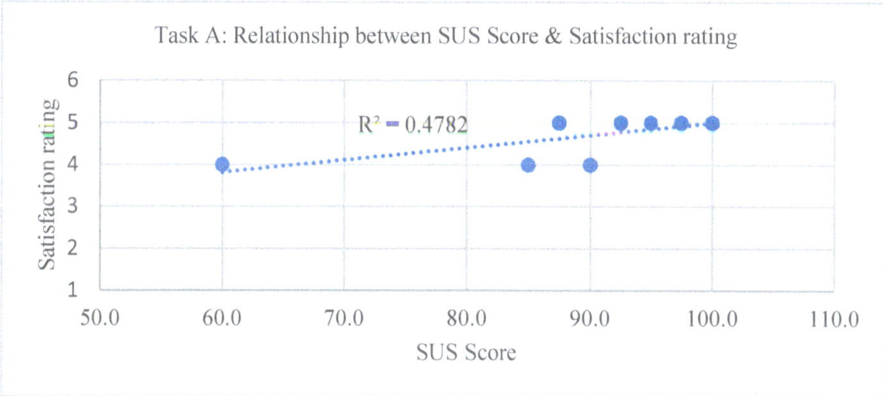

Fig. 6. Task A text search correlation of SUS score and satisfaction rating.

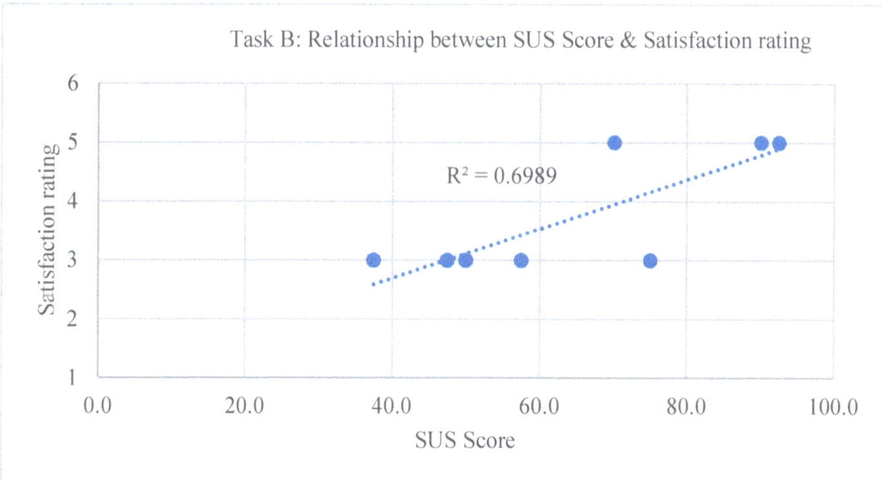

Fig. 7. Task B image search correlation of SUS score and satisfaction rating.

5 Summary of Findings

Preliminary findings do suggest that the backend technology of search determines the efficacy (and usability) of the frontend search experience. This is important because it suggests that an image-based search query may be more effective than traditional text-based queries. The data shows that, even though the self-reported SUS score and satisfaction rating are higher for text searches, there may be an incongruity when the task times and completion rates are looked at with the effectiveness determined from the correlated usability metrics of the SUS and SEQ (satisfaction rating). The incongruity suggests that even though the Task B image search was rated lower, users were able to use it more effectively to obtain more accurate results. The three summary findings I have identified from this research are the following:

1. There seems to be a strong correlation between SUS score and satisfaction rating, and this seems to be tied to task time and completion rates.
2. There seems to be a weak correlation between task time and satisfaction rating; a larger sample would provide better insight into this.
3. Even though Task B (image search) had lower SUS score and satisfaction rating, its lower variability with task time and higher completion rate suggests that the search results were more effective than Task A. This finding reflects the literature review and my hypothesis.

6 Conclusion

My research question asked whether the image search results would be more effective, but what is effective in this case? If something seems more usable—measured by task time (people can use it faster) and completion rate (tasks are completed with fewer errors)—but receives lower self-reported SUS and satisfaction scores, what can account for this incongruity? It is difficult to measure sentiment, and these findings seem to reflect the prevailing research, which suggests that people prefer to use systems they are familiar with, even if those systems are less usable. Perhaps information and system designers have to tolerate (and expect) some errors and "unusable" attributes to account for user preference.

References

1. Jing, Y., Liu, D., Kislyuk, D., Zhai, A., Xu, J., Donahue, J., Tavel, S.: Visual search at pinterest. In: Proceedings of the 21th ACM SIGKDD International Conference on Knowledge Discovery and Data Mining. pp. 1889–1898. ACM, New York (2015)
2. Yoon, J.: Searching images in daily life. Libr. Inf. Sci. Res. **33**, 269–275 (2011)
3. Westman, S.: Image users' needs and searching behaviour. In: Chair, A.G., Davies, J. (eds.) Information Retrieval: Searching in the 21st Century, pp. 63–83. Wiley (2009)
4. Brumberger, E.: Visual communication in the workplace: a survey of practice. Tech. Commun. Q. **16**, 369–395 (2007)
5. Baker, M.: Every Page Is Page One: Topic-Based Writing for Technical Communication and the Web. XML Press, Laguna Hills (2013)
6. Goker, A., Davies, J.: Information Retrieval: Searching in the 21st Century. Wiley, Hoboken (2009)
7. Yeh, T., Chang, T.-H., Miller, R.C.: Sikuli: Using GUI screenshots for search and automation. In: Proceedings of the 22nd Annual ACM Symposium on User Interface Software and Technology. pp. 183–192. ACM, New York (2009)
8. Nielsen, J.: Why You Only Need to Test with 5 Users. http://www.nngroup.com/articles/why-you-only-need-to-test-with-5-users/
9. Usability.gov: Ensure Usable Search Results. http://guidelines.usability.gov/guidelines/186

10. Nielsen, J.: Search: Visible and Simple. http://www.nngroup.com/articles/search-visible-and-simple/
11. Spool, J.M.: Are There Users Who Always Search? https://www.uie.com/articles/always_search/
12. Voorhees, E.M., Over, P., Soboroff, I.: Building better search engines by measuring search quality, pp. 22–30. IT Professional, Los Alamitos (2014)
13. Tran, T., Mathäß, T., Haase, P.: Usability of keyword-driven schema-agnostic search. In: Aroyo, L., Antoniou, G., Hyvönen, E., ten Teije, A., Stuckenschmidt, H., Cabral, L., Tudorache, T. (eds.) The Semantic Web: Research and Applications, pp. 349–364. Springer, Berlin Heidelberg (2010)
14. Choi, Y.: Effects of contextual factors on image searching on the Web. J. Am. Soc. Inf. Sci. Technol. **61**, 2011–2028 (2010)
15. Yeh, T., White, B., San Pedro, J., Katz, B., Davis, L.S.: A case for query by image and text content: searching computer help using screenshots and keywords. In: Proceedings of the 20th International Conference on World Wide Web. pp. 775–784. ACM, New York (2011)
16. Tullis, T., Albert, W.: Measuring the User Experience: Collecting, Analyzing, and Presenting Usability Metrics. Morgan Kaufmann, Burlington (2008)
17. Charney, D.: Experimental and quasi-experimental research. In: Research in Technical Communication, p. 266. Greenwood Publishing Group (2002)
18. MacNealy, M.S.: Strategies for Empirical Research in Writing. Allyn and Bacon, Boston (1999)
19. Brooke, J.: SUS-a quick and dirty usability scale. Usability Eval. Ind. **189**, 4–7 (1996)
20. Bangor, A., Kortum, P.T., Miller, J.T.: An empirical evaluation of the system usability scale. Int. J. Hum.-Comput. Interact. **24**, 574–594 (2008)
21. Lewis, J.R., Sauro, J.: The factor structure of the system usability scale. In: Kurosu, M. (ed.) Human Centered Design, pp. 94–103. Springer, Berlin Heidelberg (2009)
22. Measuring satisfaction: Beyond the usability questionnaire, http://www.userfocus.co.uk/articles/satisfaction.html
23. Nielsen, Jakob: User Satisfaction vs. Performance Metrics, http://www.nngroup.com/articles/satisfaction-vs-performance-metrics/
24. Sauro, J., Dumas, J.S.: Comparison of three one-question, post-task usability questionnaires. In: Proceedings of the SIGCHI Conference on Human Factors in Computing Systems, pp. 1599–1608. ACM, New York (2009)
25. Sauro, J., Lewis, J.R.: Correlations among prototypical usability metrics: evidence for the construct of usability. In: Proceedings of the SIGCHI Conference on Human Factors in Computing Systems, pp. 1609–1618. ACM, New York (2009)
26. Sauro, J., Lewis, J.R.: Estimating completion rates from small samples using binomial confidence intervals: comparisons and recommendations. In: Proceedings of the Human Factors and Ergonomics Society Annual Meeting, vol. 49, pp. 2100–2103 (2005)

Teaching Methodologies in Communication Design

Cátia Rijo[1(✉)], Helena Grácio[1], and Sandra Antunes[2]

[1] CIAUD, Faculty of Architecture of the University Lisbon, Rua Sá Nogueira, Pólo Universitário, Alto da Ajuda, 1349-063 Lisbon, Portugal
catia.rijo@gmail.com, ahgracio@gmail.com
[2] CIED, School of Education of Lisbon of the Polytechnic of Lisboa, Campus de Benfica do IPL, 1540-003 Lisbon, Portugal
antunes.sandra.design@gmail.com

Abstract. This paper aims to contribute for the discussion in design teaching and its consequent research context, in design research and teaching. It is intended to discuss the importance of reflecting and analyzing the *modus operandi* usually adopted and the results obtained by adopting a design process methodology in the educational field, particularly in Communication Design Education. Therefore, our main goal is to implement and propose a methodology centered and driven by practical action, as much as possible in a real context scenario, structured according to the different phases of a design project. Signed in collaborative processes, the action developed it is characterized by the interdisciplinary dialogue in formal, informal, generational and culturally plural contexts able to characterize and test and evaluate the approach to different thematic, enabling to enriches and optimize the design experience as the process of apprehension.

Keywords: Teaching methodologies · Design · Communication · Creativity

1 Introduction

Designed within a Polytechnic Higher Education institution, whose specificity is built on the affirmation of the identity of its action characterized, among others, by the importance given to experimentation, observation and integration of material and immaterial culture, knowledge, techniques and the deposed technologies in productive activities, from which derives the *greatest sensitivity to cyclical changes* [1] and the willingness to reflect on them that characterizes us - this project arises.

In this action, it was developed a focused methodology, conducted by practical action as much as possible in a real context, structured according to the different phases of a communication design project. Signed in collaborative processes, we will characterize the action developed by the interdisciplinary dialogue in formal, informal, generational and culturally plural contexts.

Considering the practices and methodologies exposed, the proposed research will be developed in search of alternative educational practices to the non-participatory model, centered in the classroom - thus contributing to the research, and the creative development, of prospective Teaching and research in design.

A.G. Ho (ed.), *Advances in Communication of Design*, Advances in Intelligent Systems and Computing 609, DOI 10.1007/978-3-319-60477-0_4

Look to our experience in the classroom and at school, as an *effective training tool* [2] and the subsequent observation of the student body, took place the proposed project.

In the context of the Visual Arts and Technologies degree, at the Lisbon School of Education, in the design field disciplines, it was established by oriented projects inter- and transdisciplinary, aiming to address matters in various spheres of design: Communication; Web; Product; Inclusive and Social partnership with the school community and its institutional partners - strengthening relations between the human and social spheres, market demands and research in an academic context; Contributing decisively to the training of the student as a researcher or professionals in this area and to our integrated/applied research in Design;

As inter space and transdisciplinary these curricular units have become a project laboratory for which different agents compete with their knowledge and methodologies of different areas of knowledge. By this way, the actors and the work tools, of interdisciplinary provenance, are dynamic, adapting themselves in accordance with the specifications that best suit the activity for which they are intended.

Evidence of the practices in its genesis include, among others, design actions such as: Nisa Residence, StoryLab, CulturEse and Sabistória.

Mediating between the different contexts of the *ideal* and the *real,* this project makes it possible to obtain *feedback* about: whether the work processes; either of the conditions of reception of the work, or of market/person expectations and the actual use of knowledge and of the product is an effect - allowing the collection of data impossible to obtain in an isolated context and the consonant to redesign the strategy of production, and our pedagogical action as teachers and Polytechnic School; or even the student as a future creative and professional entrepreneur.

To grow, build structures, sediment knowledge, reflect on work processes, generate and plan innovation strategies - germinating use value research and production.

2 Problem

The test Christopher Frayling (1993/1994) is one of the first publications to explain the term *practice-led research.* A distinction drawn by the author between the concepts of "research *for* art practice" (conducted to produce better works of art); "Research *through* art practice" (considering the practice as an integral part of the research design) and "research *into* art practice" [3] (referred to research that sets the aim of clarifying the process and creative practice).

In the academic world, several questions have been raised about how to develop a project-oriented practice capable of generating knowledge; or for the institutional processes through which knowledge can qualify as research. On this very question have reflected Biggs & Büchler. A demand whose great development was based in the United Kingdom in 1992 by the merger between polytechnics and universities and the consequent need to conduct research at the doctoral level in the areas of project practice at the time [4].

According to these authors, established in 1998 and elevated to the *council* category in 2005, is due to the United Kingdom "Arts and Humanities Research Board", the pioneering action towards the legitimization of this methodology as competent in generating knowledge when for the first time published clear criteria on what he believed would be capable of constituting research in the disciplinary areas under his control. Paradigmatic in consideration of practice as method, not only as a means, it is still the project of the University of Hertfordshire "Non-traditional Knowledge and Communication", directed the investigation around the problems arising, improving processes and creating consistency research at the level of creative and cultural industries. Projects to observe in the design of methodological assumptions based on the development of collaborative platforms as a result of this project in the prop we start.

3 Objectives

Understand that capital gains for projectual methodological practice, result from the action of a working group consisting of officers of the various aspects of the design where you try to understand that particular place of immersion of students in local communities: with its people, their culture, the particularity of its know-how, technological and empirical, their identity, their memory, but also the sent gone their future aspirations.

It is intended to analyze which valences the immersion of students in real work brings, in relation to the need to develop strategies to integrate and respond to a set of interacting dynamics, as well as the experimentation of the rhythms and daily difficulties inherent to the profession and to understand how the importance given to practice and experimentation as forms of research/knowledge; observation and consideration of local material and immaterial culture; the inherent knowledge; their techniques; the raw materials and the local technologies, still preserved in the productive activities; or even the experiences outside the classroom are facilitators and propellers of learning.

It is our intention to make use of the results obtained by this way, for the development of methodologies and strategies of teaching and research in design.

4 Methodology

Forged in the parallelism with the methodologies used in the real context, adapting them to the teaching-learning process, the present project is based on pedagogical tools drawn from the elaborated constructive reflections: either from the action of teachers and students in context real; either from surveys of students and key stakeholders in each of the proposed projects; or from the moments of formative evaluation of students and teachers - finally proposing a methodology appropriate to the reality of the current students in the courses of design.

The experience of teaching in design that it is propose is characterized by the multidisciplinary of its actors; focus on research/action; the immersion of teachers and students in real local contexts (whenever possible) or the creation of conditions as similar as possible to the real context, and consideration in the projective activity of the whole human and cultural dimension of the immersion teaching/learning context.

It is intended to integrate the practice as a tool for research and action design - *practice-based* and *practice-led* [5] - considering: that practice and reflection on the results of the practice; or the artifact created as a source of new knowledge and research; or a search capable of leading to new knowledge or new practical action; one observes in this way the knowledge about the practice or within it. A question considered inherent in the action of the designer, for whom, as an agent in the area of creative practice, the results of research often manifest as non-linguistic - often opening doors to the discovery of "something" that could not have been discovered by other means [6].

Through collaborative and participative methodologies in teaching and research in design, integrating the student in a real context: motivation and accountability is promoted as part of the process; inducing decentralizing the self-student, reinforcing the focus on the end user, in the productive contexts and also the effects of the proposed product. It is therefore invited to study and rehabilitate raw materials with peculiar characteristics; creation, rather than objects, services and experiences that can revitalize the native human community and transmit heritage and local memories - projecting the effects of the products, services and experiences conceived in relation to them.

Acting in the transdisciplinary field, with support in the methodology of Design Thinking, planning to allow students some time for the exploration/experimentation of systems by non-linear pathways (with time for discussion and reformulation of proposals throughout the pre-project phase) - practices of a sense other than the specialization of knowledge are observed - immersive experience/applied research; educational practices and alternatives to the non-participatory model centered on the individual in the classroom.

Building on the practice and driven by it, the student interacts and reacts - a methodology that encourages and exponents the creativity of the students, reaching through this way results that could not have been achieved otherwise.

Through a methodology of mixed research, it is intended the definition of a set of guidelines to support the teaching of design in the context of Higher Education: in non-interventionist terms, it is understood processes of data collection by direct observation, interviews and questionnaires; as far as interventionist research is concerned, it comprises active research based on the study of cases from different classes in the curricular units related to design.

Tasks:

- Development of partnerships and creation of briefings for the teaching of the expected contents, in a situation of immersion of students and teachers in multidisciplinary teams and in real contexts;
- Accomplishment of moments of summative evaluation, integrating: surveys to the various agents involved; Moments of self, hetero-evaluation; Reflection and analysis on the data obtained;
- Writing written reflection about the analysis of the data obtained;
- Production of pedagogical tools to support teaching and research in design;
- Growing, founding structures, sedimenting knowledge, reflecting on work processes, generating and planning innovation strategies - germinating use value research and production.

5 Results

Through this line of research, it is intended to bring together the most diverse agents in a project of responsibility and social sustainability. Through the sharing of experiences and interdisciplinary issues around design focused on social causes, we undertake reflection on the participatory management of resources [7] on the part of the agents in attendance who undertook the design of new materials, new uses for practices consecrated by time, new products or their alternative dematerialization in services, experiences and observation of effects and consequences.

Local diversity and the interaction of teachers and students in scientifically, culturally and generationally plural collaborative platforms play a very significant role from a methodological point of view: either by expanding the agents' ability to observe hypothetical future realities; or providing opportunity for reframing of theoretical issues, by confronting them with real *inputs* its consequent consideration under various levels of complexity (rational, emotional, cultural, technical and technological). This process highlights the exponentiation of speculative aptitude beyond strictly logical and analytical thinking, widening the avenues for the development of new courses of action (innovation).

Considerations about the worldview and the investigative resources derived from various branches of visual design and art and local actors, allowing conjecturally to reframe the emerging problems; observe them under different perspectives ponder and consider their various levels of complexity. In this way, conceived in observation, experience and in response to the needs and aspirations of a small human group, beyond the observation of the common man, the project was humanized. In this way - decentralizing the design processes of the designer - we will be able to focus more on the user and the context of production and use, creating added value for the project.

6 Conclusions

The proposed design teaching experience is characterized by the multidisciplinary of its actors; mainly focused on a research/action methodology; always giving preference to the immersion of teachers and students in real local contexts (whenever possible); or the creation of conditions as similar as possible to a real context scenario, as to the consideration of project activity of the whole human and cultural dimension through the immersion in a teaching/learning context.

Our intention is to integrate practice as an instrument for research and action design considering as an operative tool the methodology -practice-based and practice-led. Both practice and reflection on the results of practice, or the artifact created as a source of new knowledge and research; or a search capable of leading to new knowledge or new practical action. Where it is observed the knowledge about the practice or within it. A question considered inherent in the action of the designer, for whom, as an agent in the field of creative practice, the results of the research often manifest as non-linguistic- often opening doors to the discovery of "something" that could not have been discovered by other means. Considering the practices and methodologies exposed, the proposed research will be developed in looking forward to search new alternative educational practices [procedures] to a non-participatory model, centered in the classroom, thus contributing to the research and the creative development of approximations, as teaching and researching prospective projects in design.

The purpose of the research is to assess the importance of applying less conventional methodologies in design teaching, as well as the perception of how an immersive learning context can be a facilitating tool for learning, fostering creative thinking, generating new ideas and future paths to the [problem] solution.

In the end, we intend to assess the importance of applying the new methodologies and how this immersive learning context is a facilitator of learning and fostering creativity by providing new ideas and future paths to the problem in question.

Through our action, we join the various actors in a project of social responsibility and sustainability.

Considerations about the worldview and investigative resources arising from various fields of design, visual art and the local sectors, allows us, at every moment, circumstantially reframe emerging problems; observe them under different perspectives ponder and consider their various levels of complexity. In this way, conceived in observation, experience and in response to the needs and aspirations of a small human group, beyond the observation of the common man, the project was humanized. In this process, through immersive, is encouraged develop the product material and immaterial not to, but with the community. In this way - decentralizing the design processes of the designer - we will be able to focus more on the user and the context of production and use, creating added value for the project.

In search of alternatives to educational models focused in the classroom, along with research on teaching practices and design research, stimulates to mediation, interaction and cultural and social responsibility. Simultaneously managed -if expanded the perceptual field of the various participants; the levels of reading of material and immaterial realities in presence of individual and collective memory; of scientific knowledge; of techniques and technologies, also contributed to the consideration of several levels of meaning, as providing and propelling the development of new action.

The methodology and projectual practice reinforce the above basic principles, expanding the scope of the concept and projective practice; teaching and research in design. A reality which leads us to defend the methodological proposal presented here: founded and integrated in real context of collaborative platforms in design teaching and research. A tool facilitating the acquisition, transmission, mobilization and future implementation of research skills and collaborative work, essential to the designer as competent and responsible agent.

References

1. Urban, C.V.: Polytechnic Education - (re) positioning in the panorama of higher education in Portugal. In: VI Portuguese Congress of Sociology. Social Worlds: Knowledge and Practices. New University, Lisbon (2008)
2. Dreeben, R.: The unwritten curriculum and its relation to values. In: Traduzido por Gimeno, J., Perez, A. (eds.) La enseñanza: su teoría y su práctica. Akal, Madrid (1976, 1983)

3. Webb, J.: The logic of practice? Art, the academy, and fish out of water. Beyond Practice-Led Res. (14), 1–16 (2012). http://www.textjournal.com.au/speciss/issue14/Webb.pdf. Accessed 02 Feb 2016
4. Biggs, R., Büchler, D.: Rigor and practice-based research. Des. Issues **23**(3), 62–69 (2007). Massachusetts Institute of Technology, Massachusetts
5. Candy, L.: Practice Based Research: A Guide. University of Technology, Sydney (2006)
6. Büchler, D., Biggs, M., Sandin, G., Stah, L.: Architectural Design and the Problem of Practice-Based Research. Cadernos de Pós-Graduação em Arquitetura e Urbanismo (2008). http://www.mackenzie.br/dhtm/seer/index.php/cpgau. Accessed 02 Feb 2016

Ergonomics and Organizational Communication: Methods for Business Communication Auditing

Bruna Marina Bischoff(✉)
and Flávio Anthero Nunes Vianna dos Santos

Center of Arts, University of Santa Catarina State, Av. Madre Benvenuta, 2007, Florianópolis 88.035-901, Brazil
brunamar@gmail.com

Abstract. This research is part of the author's dissertation, which will approach the communication processes in the textile industry chain. This article proposes a literature review looking forward to establish the state of art of relevant issues regarding human factors and ergonomics, related to the communication interactions research line, supported on three theoretical axes: macroergonomics, internal organizational communication, and the process of communication auditing. This article looks forward to unravel the convergences and divergences among the authors presented, showing the discussions and conclusions, as well as a preview appreciation of methods for communication auditing focused on internal communication of organizations.

Keywords: Organizational ergonomics · Business communication · Methods for communication auditing

1 Introduction

This paper proposes the accomplishment of a literature review of relevant themes in the fields of human factors and ergonomics, which were related to this author's masters' thesis, whose object of research approaches the communication processes in the industrial textile chain, mainly among the sectors of design and production in fashion companies. In order to begin understanding these relations, it is necessary to broaden the view on the theme, starting from three main theoretical axes in increasing order of depth: first in macroergonomics, second in organizational communication, and third in methods of communicational assessment.

Any kind of interpersonal relationship tends to go through conflicting moments, and the reasons for it are as varied as possible. Differences such as age, ethnicity, language, intellectual formation, ideology, politics, religion, views on society, culture, among many others, may enrich relationships, but can just as easily lead to misunderstandings and failure to exchange information.

In one's personal or family life, good communication is essential for maintaining relationships. The same applies professionally, in which the importance of communication is also widely recognized within the work environment. It may come intuitively to

A.G. Ho (ed.), *Advances in Communication of Design*, Advances in Intelligent Systems and Computing 609, DOI 10.1007/978-3-319-60477-0_5

managers and employees or through practical experiences; either way, the potential of good communicational processes in order to provide problem solutions and achieve positive results for companies is generally acknowledged. Downs and Adrian, however, argue that many organizations fall short by maintaining the habit of only referring to the relevance of communication after mistakes have already been made [1].

With this paper, the aim is to reveal the scientific findings on this theme through bibliographic research. Firstly, it describes the methodology through which the mentioned sources are found before discussing the subject of macroergonomics by drawing a parallel between the main Brazilian and international authors that research business communication. Finally, some of the most common methods used to perform communicational assessments are indicated.

2 Methodology

The search for bibliographic sources began in the online portal of Capes[1] journals, making use of the fields "Search subject" and "Search book" with the UDESC[2] login through CAFe (Federated Academic Community) in order to allow access to a wider range of publications. Search engines such as Google Scholar, Google Books and Google in general were also utilized. For the theoretical axis of macroergonomics, fellow researchers indicated authors, such as: Itiro Iida, Lia Buarque de Macedo Guimarães, Hal W. Hendrick and Brian M. Kleiner.

For the other theoretical axes, research began in Capes' online portal by looking up internal and organizational communication and methods of communicational assessment as an only subject, but few satisfactory results were found. This made it necessary to refine the keywords by separating the themes, searching for synonyms, adding and/or subtracting some terms. Thus, from the third search on, relevant results began to appear.

When the obtained results corresponded to dissertations and articles published in journals, it was possible to compare the bibliographic references used by each of them to reach classic authors on the subject of internal and organizational communication. Therefore, a document was created, containing all the books mentioned and used in such articles, and listing the authors in alphabetical order, so as to reveal the recurrence of some scientists. With this, three important Brazilian scholars were reached, as they were cited at least twice in the aforementioned papers: Margarida Maria Krohling Kunsch, Maria Alzira Pimenta and Francisco Gaudêncio Torquato do Rego. Some books used in previous research are also referred to in this work, such as "A Textbook of Business

[1] The Periodicals Portal, of the Coordination for the Improvement of Higher Education Personnel (Capes), is a virtual library that gathers and makes available to Brazilian educational and research institutions the best of international scientific production.

[2] The University of Santa Catarina State (UDESC) has excellence in higher education working in the areas of teaching, research and extension. Currently, there are 15 thousand students distributed in 55 undergraduate courses and 42 master's and doctorates, which are offered free of charge.

Communications", by Chappell and Read, and "Administração de Produção e de Operações: Manufatura e Serviços - Uma Abordagem Estratégica"[3], by Corrêa and Corrêa.

Next, bibliographic sources related to the methods of communicational assessment came into play. The search engine used at the beginning was Google Scholar, with the phrase "methods of communicational assessment", although the results found were not satisfactory. Then the search was repeated in Google's general search engine, resulting in a total of 242,000 responses.

In an extensive preliminary analysis of the many obtained results, the University Institute of Lisbon – ISCTE Business School stood out, as the book "Assessing Organizational Communication: Strategic Communication Audits", by Downs and Adrian, was included in the bibliography of the Organizational Communication class within their Human Resources Management undergraduate program. Inspired by the title of this work, the sentence "audits organizational communication" was searched for on Google Books, leading to the book "Auditing Organizational Communication: Handbook of Research, Theory and Practice" by Hargie and Tourish. These titles, then, were selected to compose the bibliography of the second theoretical axis on the methods of communicational assessment.

3 Macroergonomics

Studies in the field of ergonomics began in the mid-twentieth century and were primarily concerned with the physical and operational aspects of human labor. Up to the present day, the objects of study in this area have expanded to cover many different aspects of human activity. Iida and Buarque explain that the service sector, in relation to transportation, leisure, among others, as well as the study of minorities like obese, elderly and physically disabled individuals, indicate the main directions in which research on ergonomics is currently underway [2].

It is possible to notice that, as work practices have evolved over the last century, science in the field of ergonomics has also developed. Previously, work required a lot of repetitive physical effort. Today, it depends mainly on cognitive aspects, that is, perception, information processing and decision making [2]:

- 1950/1960 – Physical Ergonomics: researchers acted as consultants to solve specific problems related to the interactions between humans and machines;
- 1970 – Physical Systems Ergonomics: the "environment" component was added as an influencer in the man-machine relationship and ergonomics was acknowledged as a methodology;
- 1980 – Cognitive Ergonomics: with the computerization and automation of the work, as well as the creation of the Internet, the exchange of information started to happen in a much faster pace. These factors redirected the research of ergonomics to cognitive work, related to information transmission, memorization and decision making processes;

[3] "Production and Operation Management: A Strategic Approach to Manufacturing and Services" (translation by author).

- 1990 – Macroergonomics: organizational aspects were recognized in the field of ergonomics, which started to offer constant contributions to companies, and no longer in a sporadic fashion as had happened previously.

The first three stages, which constitute the so-called microergonomics, were focused on the individual and the direct relationship with hardware, software, the environment and the work itself [3]. Moreover, the last phase has the systemic level of work as its object of research, corresponding to the beginning of the grasp on macroergonomics.

The goal of macroergonomics is to optimize the work system's design in terms of its sociotechnical system characteristics, and then carry the characteristics of the overall work system design down through to the design of individual jobs and human-machine and human-software interfaces to ensure a fully harmonized work system [3].

Iida and Buarque use the terms Macroergonomics and Organizational Ergonomics as synonyms, referring to the field that integrates entire sectors within companies, especially with regard to the design and development of projects, as well as organizations' management at a strategic level. Macroergonomics can contribute by reflecting on the level of employment, on the qualification of employees, on the organization of production, on the company's investments and its overall competitiveness, always taking socio-cultural and technological factors into account when searching for the balance between personal well-being and the resourcefulness of the organizational system [2].

4 Organizational Communication

With the establishment of the Industrial Revolution, activities previously carried out in their totality by the figure of the artisan were divided into stages of production; a logic of the segmentation of processes and fragmentation of activities that remains guiding the structure of companies and industries to this day. Corrêa and Corrêa describe some of the activities that make up an example of the contemporary manufacturing system, extending until it arrives at the consumer: it starts with the production of raw material, followed by the production of components, sub-assembly production, assembly production, product assembly, distribution, and reaches completion in retail [4].

By fragmenting tasks, the inventive minds that think about the tasks work in a fragmented manner. Those activities, once so intimately known by a single artisan, are divided into sectors, niches and islands, among many other terminologies. Chappell and Read explain that, as organizations grow, the level of specialized work also increases as the activities are divided into sectors – purchase, sales, marketing, development, production, inventory, logistics, among others – and the communication between specialists adopts its own language with the use of technical terms and jargon [5]. This way, little time is spared in peer communication and it is difficult to transmit information between people and/or sectors with different specific knowledge.

In a segmented work environment, where an activity is entirely dependent on the one before it and they are both carried out by individuals imbued with particularities that go well beyond their distinct formation, the importance of establishing an efficient communication process is revealed. Hargie and Tourish corroborate this idea when they affirm, many business problems are the product of poor communications policies [6]. In this sense, Downs and Adrian [1] and Hargie and Tourish [6] emphasize the need for periodic monitoring of the effectiveness of communication, establishing what can be called an internal benchmark system to identify healthy and problematic processes, as well as those that have the potential to develop difficulties.

At the most fundamental level, organizations need to monitor how well employees communicate because the organization's survival often depends on workers' abilities to exchange and coordinate information [1].

There is a connection between internal communication and job satisfaction, as Downs and Adrian argue, generating benefits such as increased productivity, innovation and quality of goods and services, reduction of absence, strikes and costs [1]. It is thus perceived that the quality of internal communication is directly linked to positive organizational changes.

Rego proposes a systemic model to understand the processes of organizational communication; for this, he classifies a few types of communication according to different criteria, as showcased in the following table [7] (Table 1):

According to Rego, one of the objectives of this proposal is to demonstrate that organizational effectiveness has no room for growth when the different centers, areas or sectors of the same company cannot find unity in relation to their discourses, that is when they are not inserted in the same classification. The author states, the effectiveness of the organization depends, fundamentally, on the harmonious set that is arranged, having coordination structure as a pillar for organizational communication operations [7].

Table 1. Criteria for communicational classification.

Criteria	Classification
Nature	Visual communication
	Written communication
	Oral communication
Direction	Unidirectional communication
	Reciprocal communication
	Multidirectional communication
Categories	A – happen within the organizational system
	B – happen outside the company
Dimension	Behavioral – involving one or more people in the organization
	Social – including society as well as the company
	Cybernetics – encompassing digital technologies capable of storing, processing and disseminating information
Levels	Intrapersonal – related to an individual's behavior
	Interpersonal – the relationship between people and their subjectivities
	Organizational – various situations about the group within the organization
	Technological – also refers to digital equipment and its languages
Flow	Ascending
	Descending
	Lateral

Hargie and Tourish understand that managers do not always diagnose the real value that communicational audits can bring to organizations, so auditors first need to demonstrate the benefits that this practice can bring if properly implemented [6]. The following section seeks to explore communicational assessments more deeply.

5 Methods of Communicational Assessment

Any communication audit begins by considering the complexity of the communication as a dynamic and constant process. The reality of internal communication in organizations comprises a scenario with many people involved, receiving and transmitting messages in various media formats, using the information and knowledge transmitted to perform interpretations that support decisions and attitudes that constantly feed communication currents.

When receiving any type of information, the accepting individual performs an unconscious action of filtering the obtained content, so it can be stated that the message sent is not always the one received. Downs and Adrian argue that, in audits, it is quite useful to identify how managers filter the information they receive so as to understand how decisions are made. The authors argue that a deeper comprehension of the filtering phenomenon may corroborate the understanding of how communication relates to non-communicational variables, such as costs, goals and values, because of their influence on the decision making process [1].

Downs and Adrian present the concept of auditing as merely a process of exploring, examining, monitoring, and evaluating something [1]. This understanding, when related to internal communication, refers to viewing a scenario of the present moment in the organizations. Consequently, auditors look at how the communication process relates to results for the organization, such as commitment, satisfaction, productivity, effectiveness in implementing changes and financial stability.

Hargie and Tourish believe that measuring the communicational environment in organizations is extremely difficult and, in order to achieve a certain level of accuracy, a systematic approach to building well-structured, rigorously evaluated strategies is necessary [6]. Some characteristics considered essential in any communication assessment, according to Downs and Adrian [1], are:

- **Independence:** It is not interesting for companies or individuals to conduct their own assessments since there is personal interest involved. It is indicated that independent and qualified professionals be responsible for evaluations, especially because employees feel more comfortable answering questionnaires to outsiders, out of fear or shame of letting their superiors know what they really think;
- **Professionalism:** Assessments must be carried out by trained auditors, who need to be aware of the dynamism and complexity of the communication relations so that the research achieves credibility and accuracy, as well as being prepared to deal with adverse situations, such as cases of confrontation with management teams or resistance in maintaining damaging habits;
- **Rigor:** The data collected must be used to realistically describe an organization's current scenario, using systematic area assessments of potential importance.

Evaluators should not give in to the tendency to reveal problems for which they already have models of proposed solutions, which would compromise the whole effort resulting in an unrealistic and tendentious study;

- **Quality:** The use of criteria differentiates qualified professionals from anyone capable of collecting information. This allows for a more accurate interpretation of the data obtained, reducing the possibility of generalized assumptions and other distortions. Observations must be placed in their historical, organizational, industrial, or economic context;
- **Customization:** Both the evaluation methodology and the criteria for this must be specifically developed for each organization, unless an adaptation process is carried out based on previous assessments. The use of "questionnaire templates" found publicly on the Internet is not recommended. It is indicated that several methods of collecting data are used in a joint fashion, to customize the analysis according to the company;
- **Actuality:** The evaluation process is usually long, however, results reveal a scenario about a specific moment of the organization. Knowing that communication patterns are not static, periodic assessments are necessary so that the information generated remains current.

Downs and Adrian further explain that audits work to clarify the contrast between reality and the ideal workplace that would exist if the suggested results were implemented by organizations [1]. To do so, some criteria must be established to govern judgment and propose adjustments.

These criteria may be based on (1) standards generated by the organization or a professional organization such as the Occupational Safety and Health Administration (OSHA), (2) pet theories of the auditors, (3) purposes announced by management, (4) benchmark comparisons with other successful organizations, (5) internal comparisons over time, (6) current popular notions of management of leadership, or (7) objectives - and sometimes the performance levels of competitors - against which the organization can be judged [1].

Hargie and Tourish present eight techniques of data collection and methodologies for communicational assessment in detail [6]. Two of these, considered of great importance and practical applicability, are presented in the following section.

5.1 Assessment Methods Using Forms

Unlike the questionnaire method, which happens upon reflection on situations, patterns or tendencies occurred in the past, forms take place in the present moment, soon after the occurrence of communication. The two methods presented here use the technology of forms; the first one requires that the individual whose attitudes are being observed be the one responsible for completing the information, while in the second method an evaluator provides the information after interacting directly with the observed subject.

Journal Entries. This method consists of collecting subjective information such as behavior, experiences and social interactions in a defined period of time, through the written report of selected individuals. Although it may be considered more complex than a questionnaire, it is highly efficient due to the subjectivity factor of the answers.

It is open to provide data regarding the history of interpersonal relations, frequency, subject and duration of communications, however most reports tend to refer to formal meetings and contacts, to the detriment of informal dialogs, which can lead to data analysis failures.

The advantages of this method are: it is a natural and familiar activity, generally well accepted by the participants; the ability to inform the source, recipient, nature and effectiveness in each individual communication; it does not require the presence of the auditor at the time of communication; it provides details and personal data of those who are immersed in the reality of communication.

On the other hand, the disadvantages include: lack of control over the quality and truthfulness of what is answered; reality can be masked by the personal opinion of the respondent; high dropout rates, according to the duration of the observation period; the presence of the evaluator may affect the outcome of the responses.

Secret Audit. With this method, the researcher or a trained observer acts as a client of the evaluated company to make use of the products or services provided, without the employees realizing that they are being observed and assessed. One of the great advantages of this method is that it can be applied in several segments. In general, the process that follows this technique is the filling of forms after the observation.

6 Final Considerations

This paper aimed to address the topic of internal organizational communication and methods of communicational assessment. A bibliographic research was developed gathering concepts and opinions of authors and researchers referenced in Brazil and internationally. Through the discussed subjects, it was possible to perceive the importance of maintaining effective communication processes in the reality of contemporary companies, as most are departmentalized in several sectors.

Attention was also given to the difficulty in carrying out measurements regarding the communicational environment in organizations, and so systematic, periodic and well founded surveys and evaluations are required, taking into account specific criteria such as independence, professionalism, rigor, quality, customization and actuality. Two methods used in communicational audits with wide practical applicability were also exposed. With this, it is possible to affirm that the objectives of this work were reached.

References

1. Downs, C.W., Adrian, A.D.: Assessing Organizational Communication: Strategic Communication Audits. The Guilford Press, New York (2004)
2. Iida, I., Buarque, L.: Ergonomia: Projeto e Produção. Blucher, São Paulo (2016)
3. Hendrick, H.W., Kleiner, B.M.: Macroergonomics: Theory, Methods, and Applications. CRC Press, Boca Raton (2005)

4. Corrêa, H.L., Corrêa, C.A.: Administração de produção e de operações: manufatura e serviços: uma abordagem estratégica. Atlas, São Paulo (2008)
5. Chappell, R.T., Read, W.L.: Comunicação Interna da Empresa Moderna. Forum, Rio de Janeiro (1973)
6. Hargie, O., Tourish, D.: Auditing Organizational Communication: Handbook of Research Theory and Practice. Routledge, New York (2009)
7. Rego, F.: Comunicação empresarial, comunicação institucional: conceitos, estratégias, sistemas, estrutura, planejamento e técnicas. Summus, São Paulo (1986)

Trust Building Privacy Preferences for Young Adults Visiting Political Campaign Websites

Timothy Richards(✉) and A. Ant Ozok

University of Maryland Baltimore County, Baltimore, MD, USA
{timrichl, ozok}@umbc.edu

Abstract. Websites have been used by political campaigns for nearly a decade and a half and the literature has adequately covered human factors aspects of political campaign websites with the help of usability, task analysis, and content analysis studies. However, there is a gap in the literature concerning the relationship between campaign website visitor privacy preferences and trust in the candidate or campaign. This paper presents the idea that user preferences regarding political websites and privacy policies can inform the design process of such websites and impact user trust. The paper provides explicit recommendations for improving campaign website design with the ultimate goal of facilitating trust between young adults and those using a campaign website as a part of their bid for public office. It uses mixed methods to ascertain the privacy preferences among young adult political campaign website users and identifies several factors that have an impact on trust.

Keywords: Human factors · Privacy · Trust · Campaign websites · Political communication · Mixed methods

1 Introduction

Prior research has shown that young adults are a growing opportunity for political outreach via the Internet but that their user experience on sites used for political endeavors may not match their needs or expectations [1]. Reports from the US Department of Education use the age range between eighteen and twenty-four when analyzing education and young adults [2]. Pew uses the age range of eighteen to twenty-nine when discussing technology trends in social media and politics [3]. For this paper young adults are considered to be between eighteen and twenty-five.

The disciplines which developed e-commerce over the past several decades have provided technology, research, and ideology that government agencies and politicians are adopting at an ever-increasing pace. The overarching label given to such endeavors is E-government [4]. Related to e-government is e-democracy or digital democracy, the use of the World Wide Web and Internet by politicians, political campaigns, political organizations, and special interest groups [5, 6]. Opinions abound regarding the effectiveness of the Internet in the advancement of democracy, yet Bimber and Davis [7] point out that there exists little scientifically collected data regarding this phenomenon.

The purpose of this study is to explore the maturation of the Internet and World Wide Web in political campaigns within the United States from the privacy perspective in order

A.G. Ho (ed.), *Advances in Communication of Design*, Advances in Intelligent Systems and Computing 609, DOI 10.1007/978-3-319-60477-0_6

to understand how these factors affect young adult trust in political campaign websites. A framework is presented that models the campaign website privacy policy preferences of young adults. This framework is built by deploying an online survey that collects quantitative data and a series of interviews that obtain qualitative data. The information gleaned from data collection and analysis contributes a privacy model for campaign websites that positively influences young adults' trust in the website and candidate.

2 Literature Review

2.1 Political Communication and Campaign Websites

Blumler and Kavanagh [8] described what they labeled the third age of political communication, which is characterized by an increase in "cyber politics." Over a decade later this predicted age has arrived and the use of the World Wide Web (WWW) and the Internet within the political scene has matured consistently with each election [9]. The Internet as a new media used extensively for campaigns is due in part to its lower costs and flexibility for campaign control of message [10]. There is a body of literature that focuses on elections around the world and the impact of the Internet and the World Wide Web upon these campaigns and races [11]. A survey conducted after the 2005 UK elections found that when comparing the use of the Internet and websites for campaign purposes by young adults and older adults, young adults had a higher level of use frequency or had engaged in a broader array of digital activities [12].

Website use by political campaigns debuted for presidential elections in the 1992 Bush-Clinton race, and by the mid-1990s national parties and most congressional candidates had implemented a web presence, with national campaigns relying more heavily on web sites by the turn of the millennium [10]. In their study of how and when campaigns use web sites, Druckman et al. [13] looked at the issues of interactivity, data, and personalization as features and opportunities for campaign sites. Yet, their review of 2002 and 2004 congressional race websites is lacking a human factors component. A study of U.S. Presidential campaign websites performed by Benoit and Benoit [14] looked at both the content of said sites as well as some usability factors. In the study, an analysis of Bush and Gore websites in 2000 was performed just before the election in which the factors of identification, navigation, readability, irritability, information accessibility, interest level, information breadth and depth, adaptation to audience, and interactivity were examined and detailed to provide a framework whereby political sites of the future could be evaluated [14]. Modern websites provide an ever-increasingly rich user experience on political websites due to increased interactivity, social media and other Web 2.0 features [15].

2.2 Human Factors

Trust. Walczuch and Lundgren [16] conducted a survey to identify the major psychological factors that affect trust in e-commerce conducted via the web. Their analysis revealed that of the factors studied (perception-based, personality-based, knowledge-based, attitudinal, and experience-based factors), trust had a strong link to perception-based and cognitive factors [16]. In another study that examined trust in e-commerce, Cases, Fournier [17] look at trust and acceptance in e-mail campaigns as they relate to purchase intention. This study revealed that a direct link between attitudes regarding e-mail campaigns and websites does not exist but that in their own right trust and acceptance are factors that can be applied to both [17]. Additionally, Lindgaard et al. [18] performed several experiments that show a budding correlation between visual appeal and trustworthiness of homepages. These studies show that this ethic of trust carries into the design space of interfaces and web design. A political candidate and campaign seek to garner trust from their constituents.

Privacy. Privacy is an ethical concern that can have an impact on user satisfaction and trust and can evoke strong emotions from users. This study researches privacy, data collection, and data sharing within the context of political campaign websites. It examines users' preferences regarding their online privacy and privacy policies. The idea of privacy is a construct rooted in the social and cultural fabric of a people and, therefore, the rights, challenges, and definitions are subject to change over time [19]. This means that different sociological cohorts will have differing views on privacy. This dissertation focuses on the current cohort aged eighteen to twenty-five. A recent definition of privacy is "the quality of or state of being apart from company or observation" and a "freedom from unauthorized intrusion" [20].

Privacy Policies. Privacy policies are used by many organization but have an incredibly long way to go before they are user centric [21]. Such documents should be accessible, readable, clear, and not just placed into existence to provide legal cover [21]. The survey and interview guide used in this research project ask questions that examine user expectations and wishes regarding the accessibility, readability, and purpose of privacy policies on political campaign websites. Hong and Thong [22] performed a broad review of the literature regarding what they label "Internet Privacy Concerns" in which they detail nearly two decades of privacy research that frequently examine data collection and awareness as significant variables. More recently, the privacy paradox, the difference between what users say about their privacy preferences and their actual actions or behaviors, has become interesting to researchers [23]. This paper extends the prior research by examining these variables coupled with privacy policy readability and visibility as human factors that impact trust on campaign websites.

2.3 Research Questions

Research Question One (RQ1). *What is the level of knowledge among the users of political websites regarding the records of donations given to a political candidate via a website being public information by law?*

The Federal Election Commission was founded in 1975 to provide oversight and governance for the election process of federal candidates – particularly as it relates to issues of finance. Many state governments also house agencies or bureaus responsible for monitoring not only state elections but also the financing of campaigns. The laws that govern campaign finance require that most donations made by individuals or organizations be made public – a publication which in many cases is made available via a searchable web-based database. Hunter [24] indicates that the process of e-campaigning has ushered into existence a new privacy issue with regards to democracy. Gardner [25] attempted to construct a new framework for considering policies which evaluate anonymity and public record and their influence on the democratic process. This study considers the awareness of website users regarding existing regulation regarding donations they may make. While website visitors may be concerned with issues of privacy and anonymity, federal and state laws require some data to be shared and made public. Regardless of user preference, these laws exist and the campaign websites must comply with data collection and sharing requirements. How privacy measures are implemented on a campaign website is affected by these laws and visitors' knowledge regarding their legal rights and responsibilities may be unknown to them.

Research Question Two (RQ2*).* *Do political website visitors believe that they can trust the candidate more if the website has a clearly written privacy policy when compared to campaign websites which do not?*

RQ2 centers on the issue of privacy perceptions on political websites as a human factors design concern. Privacy has recently seen an increase of attention both in the media and research and both theoretical and domain-specific research can be found in the HCC literature. Parayitam et al. [26] discovered that of all company policies on all e-commerce websites examined in their study, more attention was given to firm's privacy policy than any other topic. Proctor et al. [27] examine the usability of web privacy policies and determine that amid all of the attention given to these policies readability and general usability is still lacking. It is this research question's objective to ascertain the extent to which the perception of privacy during the use of a political website matters with regard to the readability of the campaign website's privacy policy.

Research Question Three (RQ3). *Do political website visitors believe that they can trust the candidate more if the website has a clearly visible privacy policy when compared to campaign websites which do not?*

The work of Flavián et al. [28] reveals that visitor trust of a website has a relationship to a site's "perceived usability." Green and Pearson [29] performed a study built on Lee and Kozar's work which formulates an B2C e-commerce acceptance model as it relates to the HCI and e-commerce websites that finds that user perceptions play an important role in the user's acceptance of the site. Furthermore, Walczuch and Lundgren [16] write that perception, experience, and knowledge have an influence on credibility in e-commerce. These studies exhibit that the notion of trust is a key component of a successful e-commerce website and that perceptions play a key role in credibility. RQ3 examines the visibility of a privacy policy as it affects the young adult users' perception of the candidate.

3 Methodology

3.1 Survey

Surveys are a common data collection instrument in political science, political communication, and particularly human factors and user satisfaction research [30]. Adams and Cox [31] believe that surveys are an excellent way of discovering what is important or trivial to the participant or user. Holbert and Bucy [32] point out that "survey methods remain a dominant methodological tool for political communication analysis" [32].

The survey used in this study is a cross-section survey that is offered within a single point in time and its goal is not to measure changes over time but rather to present a "snapshot" in time [33]. Hoffman and Young [34] list voting intention and behavior, political knowledge, and media use as the three key constructs that political communication surveys measure. Attention is given to measuring political knowledge in this survey. Multiple questions on the survey and interview guide obtain this data.

Following the initial disclosure and consent form, the participant is provided with instructions for completing the survey. The first part of the survey consists of the questions related to the research questions. Questions regarding demographics and political activity comprise the latter part of the survey.

The survey is composed of closed-ended questions. Question types include multiple choice, multiple selection, rankings, and short answer. Several questions are in multiple parts or are designed to elicit responses regarding several concepts. This may take the form of questions that ask participants to "select all that apply" or to rate or rank the given options. In addition to the multiple choice, ranking, and open-ended questions the use of questions that ask the participate to select their response using a five-point Likert scale. The five-point scale provides an opportunity for the participant to express their response based on the extent to which they agree with a given question [30].

The researcher recruited participants by sending an email to computer science, information systems, and information technology professors at the two participating institutions. The professors contacted by the researcher decided if they would pass the invitation to participate to their students and they determined if any extra credit would be given to their students who participate in the study. All participants were currently enrolled college students and while their majors vary, all participants were enrolled in an undergraduate, computer-related general education or major course at one of the participating institutions. Some of these courses may discuss privacy within the curriculum. In effort to mitigate the opportunity for bias by limiting the time professors had to discuss topics such as ethics and privacy, data was collected during the first few weeks of the term at both institutions.

Study participants completed the survey using the web-based tool offered by Survey Monkey (www.surveymonkey.com). Access to the survey was provided via a link provided in an email forwarded by the participants' professor or posted in a Learning Management System (e.g. BlackBoard). Survey data collection took place during the first weeks of the spring 2016 semester. Some professors that were asked to invite their students to participate declined, others marketed the study and did not offer any benefit to students, and others provided extra credit to students that participated in the survey. The survey response window was four weeks in length and offered early in the term. Approximately one third of the responses needed to be removed from the dataset prior to analysis because the participants were self-selected and fell outside of the required age range or neglected to state their age in the survey. This was due to user error or due to the fact that college students who desired extra credit in their course took the survey even though they did not qualify as study participants.

3.2 Interviews

The second data collection method for this research study used semi-structured interviews. The purpose of using semi-structured interviews in this study was to obtain a set of information from the participants while allowing for the flexibility of open-ended questions and follow-up questions if necessary [35]. Each of the research questions has one or more interview questions associated with it. The interview allows the participant to express his or her feelings and thoughts regarding privacy personalization on political campaign websites. A major benefit of the interview is the open-ended nature of the questions. The open-ended questions present the participant not only the opportunity to state their preference in response to the question, but also to solicit the rational for their response by following up with a second question, "why?"

The first questions in the interview collect data to answer the research questions and the questions at the end of the survey collect demographic data. The interview consisted of primarily open-ended questions that ask the participant to describe their thoughts, feelings, wishes, or observations regarding campaigns, privacy, websites, and data.

Participants for the interviews were recruited at the same two post-secondary institutions as survey participants. Flyers were posted in multiple buildings at both campuses near study areas, high-traffic areas, and classrooms. Participants were offered compensation of twenty dollars for a thirty minute interview consisting of twenty-four questions. Participants were given the option of interviewing at either of the two campuses that recruitment took place. Subjects notified the researcher of their interest in participation by email and a mutually agreed upon time to conduct the interview was selected. The interviews took place in a private office space with no third party present and the participants' responses were separate from their identities to ensure anonymity in the results. Participants were provided an informed consent form, the interview process was explained, and participants were asked to sign the form if they were interested in proceeding. After the interview was conducted each participant was compensated in the amount of twenty dollars in cash. Most interviews required ten to twenty minutes including the greeting, informed consent review, interview, and compensation. The in-person interview recruitment process yielded ten participants. Saturation was achieved and triangulation between the data collection methods yielded plentiful results.

4 Results and Discussion

This section presents the analysis of the data collected using the online survey (N = 209) and the in-person interviews (N = 10). The survey collected quantitative data that was analyzed and the interviews collected qualitative data and which was coded. The results of both analyses were then compared and contrasted to triangulate findings. A discussion section follows the analysis for each research questions.

4.1 Demographics

The survey analysis contains data from 209 survey takers between the ages of eighteen and twenty-five with a mean age of 20.94. Sixty-four percent of survey takers report being male, thirty-five percent report being female, and one percent provided no response. The three most frequent race and ethnicity identifications were Caucasian Non-Hispanic, African American, and Asian. Eighty-five of these participants were enrolled in a public two-year institution and the remaining fifteen percent were enrolled in a public four-year institution. Of the 209 participants, thirty-seven percent report having a computer-related major; forty-eight percent report having a non-computer major; and fifteen percent reported that they are general studies, indicated they are undecided, or provided no response. Participants were of a class levels from freshman to senior.

Ten interviews were conducted with four participants being female and six participants being male. The participants were all students at a four-year institution and the sample did not include freshmen. The mean age of the participants is 21.6 and the mode is 22 and the participants reported being Caucasian Non-Hispanic, Asian, or reported multiple races or ethnicities.

4.2 Research Question One Results and Discussion

Results. Amajority (79%) of survey participants rated privacy as an important, somewhat important, or very important factor when discussing campaign websites and campaign contributions. Interestingly, while survey participants considered privacy a concern when making donations on a campaign website, a large segment of survey takers were unaware of the record keeping practices and public government databases related to campaign donation data. Fewer than fifty percent of survey participants indicated that prior to their involvement in this study they knew that government entities record campaign donor information. That number is twenty percent when subjects responded to questions asking if they knew that donor data collected by government agencies are public via a database and/or website.

Interview participants revealed a moderate knowledge regarding what donation data is made public by the government and how it is released. Half of interviews indicated some level of knowledge while the other half of responses categorically denied having any awareness. Most of the interviewees provided one to three word responses like “absolutely” (P3), “I was not” (P4), or “yeah” (P1). A few respondents exemplify the lack of awareness of young adults regarding this topic. Furthermore, these same responses reveal that the conscious thought process of young adults is not focused on the government’s collection and sharing of donation data.

Discussion. Hunter [24] reminds that digital technology use in campaigning has brought relatively new privacy concerns to the table when considering democracy. Participants from both data collection methods report a general unawareness regarding the data collected by political websites when donations are made. Thus, the answer to this research question is that political campaign website users have very little knowledge regarding the data collected from donors and made publicly available. Just as Gardner [25] assembled a framework for considering policies which anonymity and public record and their influence on the democratic process, so campaign designers must decide how they wish to use their site and what perception they want their site to receive from young adults.

4.3 Research Question Two Results and Discussion

Results. In response to the fifty-four percent of participants indicated that they agree or strongly agree that they could trust a candidate when the candidate discloses his or her policies. Fifty-two percent of survey takers said that a privacy policy should contain an even balance of "legal language when compared to common language" and fifty percent were of the opinion that a campaign website privacy policy should "protect the candidate and visitor equally.

While sixty-nine percent of survey takers agree or strongly agree that reading and understanding a website's privacy policy is essential only fourteen percent of the survey participants indicated that they read website policies on every or most sites that they visit. In contrast, thirty-seven percent of participants shared that they read website policies occasionally and forty-nine percent reported that they read these policies very rarely or have never read one.

The interview asked participants to explain their perceptions regarding the relationship between the readability of privacy policies and their trust of the candidate. Four participants responded by clearly and concisely sharing their "trust" of the candidate if the privacy policy was hard to read – each remarked their trust would be reduced or low. Others described their feelings or thoughts about hard-to-read privacy policies as "red flags" or convoluted (P1), "as bad as not having it as all" (P3), and as "not being considerate of us" (P10). Participant six clearly articulated advice for campaigns and developers by sharing that he/she things that "that it should be easier to read, that way a wider range of voters would be able to understand it." Participants saw a complicated, legalese-oriented privacy policy as a negative factor in the relationship between the candidate the voter.

Discussion. While studying trust in e-commerce, Walczuch and Lundgren [16] found that the psychological factor of perception plays a significant role in trust. The young adults studied in this study stated that a hard to read privacy policy has an impact on their trust of the candidate. This finding also contributes to the HCI literature because it confirms Furnell and Phippen's [21] general recommendations for privacy policies by applying them within the political website and new media space. Such documents should be accessible, readable, clear, and not just placed into existence to provide legal cover [21]. While young adults believe that a privacy policy should protect both the organization and the site visitor, designers must clearly define what they mean by the terms privacy and anonymity in language that the site users can understand. Campaigns can increase trust among young adult website visitors by paying careful attention to the crafting of their privacy policy.

4.4 Research Question Three Results and Discussion

Results. Of the survey takers, seventy-eight percent believe that campaign websites should have a privacy policy and eighteen percent are unsure if they believe that campaign websites should possess a privacy policy. Interview questions asked participants to describe their feelings about and trust in a candidate based on the visibility of their campaign website's privacy policy and their ability to locate the aforementioned privacy policy. Responses to the question "How does your ability to locate a candidate's privacy policy affect your trust of a candidate? Why?" consistently reveal that privacy policies should be visible or the trust of the participant will erode. Six of the ten interviewees shared that they felt that not having the policy clearly visible suggests that the candidate may be hiding something. Participant three went as far as to express the opinion that "generally, there's this image of politicians are all liars and thieves and I think that's actually generally true." In trying to explain how important full disclosure, including privacy policies, is to the public, another participant referenced President Obama's lack transparency regarding documentation during his campaign as an example of how not being up front and clearly presenting documentation can affect public trust in the campaign (P7). Participant eight noted that they would not feel comfortable donating on the candidate website if the privacy policy is not visible but that that would not affect their vote and participant nine stated that "[the site] would be a pretty weak…political website."

Discussion. The results of this study answer the research question in the affirmative. By making navigation to the privacy policy very obvious, designers can increase confidence in the candidate. These findings align with the work of Furnell and Phippen [21] whose research found that organizational privacy policies should be accessible. To make a campaign website privacy policy very easy to find for young adults and to increase trust designers can place links to the policy near data collection points on the website. The designers can also make navigation to the privacy policy a part of the entry to the home page and place a link in the more traditional locations (i.e. near the bottom of the web pages).

5 Conclusion

5.1 Implications

Young adults experience the privacy paradox when visiting political campaign websites. Young adults have concerns about the privacy implications of data collection on political campaign websites. As such, campaigns should make privacy policies highly visible and clearly written. They also should use an opt-in data collection model when designing for young adults. These steps can mitigate mistrust among young adults and present a more positive digital presence for political candidates.

5.2 Limitations

This research has three primary limitations – a limited demographic, sample representativeness, and external political or media influences.

Because this study examined young adults between the ages of eighteen and twenty five, the privacy and personalization preferences of adults outside this range were not captured. Adults older than twenty-six may have different preferences from their younger counterparts. Furthermore, because the participants resided in or near Baltimore, MD, the preferences of those from other metro areas or regions were not accounted for in the results.

Because all participants were enrolled in a post-secondary institution a limitation exists. Young adults not enrolled in college were not represented in the sample. According to the US Department of Education 40% of young adults between the ages of eighteen and twenty-four were enrolled in college in the US in 2014 [2]. A large percent of young adults in the US were not accounted for in this research project. The excluded individuals may provide data that would provide different preference results.

Because data was collected during a national Presidential primary campaign participants may have had more exposure to political advertising or political news thus creating bias that could have affected their responses to the survey. This limitation means that the participants reported preferences may be different than their responses might have been during a non-election or local election year.

5.3 Future Research

Several additional demographics could be studied in the future as the privacy and personalization preferences of different individuals could provide additional insights for campaigns and designers. Recommended demographic variables include age cohort, metro area, US region, socio-economic statues, and education level.

Since there are variations in communication dynamics when comparing individual conversations to group conversations the research in this dissertation could be extended by utilizing focus groups for data collection. Compiling responses to select questions from the survey and interview guides then comparing them to responses between private may prove interesting. This additional research may provide understanding as to how removing a semblance of privacy and, certainly, anonymity may bias conversations regarding privacy, politics, and web-behavior.

References

1. Xenos, M., Bennett, W.L.: The disconnection in online politics: the youth political web sphere and US election sites, 2002–2004. Inf. Commun. Soc. **10**(4), 443–464 (2007)
2. Digest of Education Statistics. (2015). https://nces.ed.gov/programs/digest/d15/tables/dt15_302.60.asp?current=yes. Accessed 19 Jan 2017
3. Smith, A.: Digital politics: pew research findings on technology and campaign 2012. 2013. http://www.pewinternet.org/2013/02/20/digital-politics-pew-research-findings-on-technology-and-campaign-2012/. Accessed 13 Feb 2017
4. Turban, E., et al.: Electronic Commerce: A Managerial and Social Perspective. Pearson, Upper Saddle River (2012)
5. Chadwick, A.: Web 2.0: New Challenges for the Study of e-democracy in an Era of Information Exuberance. In: Coleman, S., Shane, P.M. (eds.) Connecting Democracy: Online Consultation and the Flow of Political Communication, pp. 45–73. MIT Press, Cambridge (2012)
6. Hindman, M.: The Myth of Digital Democracy. Princeton University Press, Princeton (2009)
7. Bimber, B., Davis, R.: Campaigning Online: The Internet in U.S. Elections. Oxford University Press, New York (2003)
8. Blumler, J., Kavanagh, D.: The thrid age of political communication. Polit. Commun. **16**(3), 209–230 (1999)
9. Panagopolous, C.: Technology and the modern political campaign. In: Panagopoulos, C. (ed.) Politicking Online: The Transformation of Election Campaign Communications, pp. 1–17. Rutgers University Press, New Brunswick (2009)
10. Davis, R., et al.: The internet in u.s. election campaigns. In: Chadwick, A., Howard, P.N. (eds.) The Routledge Handbook of Internet Politics, pp. 13–24. Routledge, New York (2009)
11. Coleman, S., Price, V.: Democracy, distance, and reach. In: Coleman, S., Shane, P.M. (eds.) Connecting Democracy: Online Consultation and the Flow of Political Communication, pp. 23–43. MIT Press, Cambridge (2012)
12. Mesch, G.S., Coleman, S.: New media and new voters. In: Loader, B.D. (ed.) Young Citizens in the Digital Age, pp. 33–47. Routledge, New York (2007)
13. Druckman, J.N., Kifer, M.J., Parkin, M.: Development of candidate web sites. In: Panagopolous, C. (ed.) Politicking Online, pp. 21–47. Rutgers University Press, New Brunswick (2009)
14. Benoit, P.J., Benoit, W.L.: Criteria for evaluating political webpages. South. Commun. J. **70**(3), 230–247 (2005)
15. Chadwick, A., Howard, P.N.: Introduction. In: Chadwick, A., Howard, P.N. (eds.) Routledge Handbook of Internet Politics, pp. 1–9. Routledge, New York (2009)
16. Walczuch, R., Lundgren, H.: Psychological antecedents of institution-based consumer trust in e-retailing. Inf. Manag. **42**(1), 159–177 (2004)
17. Cases, A.-S., et al.: Web site spill over to email campaigns: the role of privacy, trust and shoppers' attitudes. J. Bus. Res. **63**(9–10), 993–999 (2010)
18. Lindgaard, G., et al.: An exploration of relations between visual appeal, trustworthiness and perceived usability of homepages. ACM Trans. Comput.-Hum. Interact. **18**(1), 1–30 (2011)
19. Spiekermann, S.: The challenges of privacy by design. Commun. ACM **55**(7), 38 (2012)
20. Merriam-Webster. Privacy. (2017). http://www.merriam-webster.com/dictionary/privacy. Accessed 20 Feb 2017
21. Furnell, S., Phippen, A.: Online privacy: a matter of policy? Comput. Fraud Secur. **2012**(8), 12–18 (2012)

22. Hong, W., Thong, J.Y.L.: Internet privacy concerns: an integrated conceptualization and four empirical studies. MIS Q. **37**(1), 275–298 (2013)
23. Brinson, N.H., Eastin, M.S.: Juxtaposing the persuasion knowledge model and privacy paradox: an experimental look at advertising personalization, public policy and public understanding. Cyberpsychol.: J. Psychosoc. Res. Cyberspace **10**(1) (2016)
24. Hunter, C.D.: Political privacy and online politics: how e-campaigning threatens voter privacy. First Mon. **7**(2) (2002)
25. Gardner, J.A.: Anonymity and democratic citizenship. William Mary Bill Rights J. **19**(4), 927–957 (2011)
26. Parayitam, S., Desai, K.J., Desai, M.J.: E-commerce policies, customer privacy and customer confidence. XIMB J. Manag. **5**(2), 45–58 (2008)
27. Proctor, R.W., Ali, M.A., Vu, K.-P.L.: Examining usability of web privacy policies. Int. J. Hum.-Comput. Interact. **24**(3), 307–328 (2008)
28. Flavián, C., Guinalíu, M., Gurrea, R.: The role played by perceived usability, satisfaction and consumer trust on website loyalty. Inf. Manag. **43**(1), 1–14 (2006)
29. Green, D.T., Pearson, J.M.: Integrating website usability with the electronic commerce acceptance model. Behav. Inf. Technol. **30**(2), 181–199 (2011)
30. Ozok, A.A.: Survey design and implementation in HCI. In: Sears, A., Jacko, J.A. (eds.) The Human-Computer Interaction Handbook, pp. 1151–1169. Lawrence Erlbaum Associates, New York (2008)
31. Adams, A., Cox, A.L.: Questionaires, in-depth interviews, and focus groups. In: Cairns, P., Cox, A.L. (eds.) Research Methods for Human-Computer Interaction, pp. 17–34. Cambridge University Press, New York (2008)
32. Holbert, R.L., Bucy, E.P.: Advancing methods and measurement. In: Budy, E.P., Holbert, R.L. (eds.) The Sourcebook for Political Communciation Research: Methods, Measures, and Analytical Techniques, pp. 3–15 Routledge, New York (2011)
33. Kenski, K., Gottfried, J.A., Jamieson, K.H.: The rolling cross-section. In: Budy, E.P., Holbert, R.L. (eds.) The Sourcebook for Political Communciation Research: Methods, Measures, and Analytical Techniques, pp. 34–54. Routledge, New York (2011)
34. Hoffman, L.H., Young, D.G.: Political communication survey research. In: Budy, E.P., Holbert, R.L. (eds.) The Sourcebook for Political Communciation Research: Methods, Measures, and Analytical Techniques, pp. 55–77. Routledge, New York (2011)
35. Merriam, S.B.: Qualitative Research: A Guide to Design and Implementation. Jossey-Boss, San Fransisco (2009)

Analysis of Communicational Design and Information on University Websites

Lívia Inglesis Barcellos(✉), Ekaterina Emmanuil Inglesis Barcellos, and Galdenoro Botura Jr.

São Paulo State University, Av. Engenheiro Luiz Edmundo Carrijo Coube, no 14-01, Bauru, SP 17033-360, Brazil
liviaibarcellos@gmail.com, ekaterina@faac.unesp.br, galdenoro@sorocaba.unesp.br

Abstract. The Universities' websites get each day more importance by acting as a communication channel that offers the most different information, reaching users in large scale. Websites are usually the first contact with the institution, or source of relevant data. Its value goes beyond the initial intention and purpose, becoming a vital part of the University's personality. This article presents the outcome of a research made with 100 persons, where 85% having the most basic level of instructions, and 15% having Ph.D. title. The research sized people's first reactions by accessing websites of the five biggest and most renowned Universities around the world, according to international rankings. Data analysis was focused on identifying the support to people's expectations and needs, besides the facility of use, the interest aroused and respect to user's time, not mattering user's experience on using internet. The adoption of an unprecedented methodology that finds points where the interface designers must act to improve the usability and the results obtained showed the existence of a big field to be improved by websites designers when the focus of this communication channel is the user.

Keywords: Visual communication and design · Websites · Universities · ICT (Information and Communications Technology)

1 Introduction

The concept of communicability [1, 2], determinant on knowledge's society, is directly related to the cyberspace's universe. The group of new communication's technologies, improved on XXI century, brought a significant impact on the generation, interconnection and sharing of information, besides its flexibility and independence in relation to how is created and stored, allowing different visions of the same content.

With the spread of Internet use, the use of websites as a channel for communication and dissemination of information has spread throughout the world and the variety of tools and collaborative resources have facilitated the creation of a structure for the diffusion and expansion of knowledge [3]. The amplification of information empowered by the management of this technology of information and communication has profoundly modified the notion of time and space, radically shortening the path

A.G. Ho (ed.), *Advances in Communication of Design*, Advances in Intelligent Systems and Computing 609, DOI 10.1007/978-3-319-60477-0_7

necessary to perform tasks and, also has transformed the form of communication of organizations and their transactions overcoming temporal and geographical barriers [3].

The organizations began to feel the need to be present in this space and the Websites were the chosen vehicle. However, defining the characteristics of these web-sites is far from a simple task, since it involves the image of the institution, the information and messages to be transmitted, the expectations and wishes of those who access them, post-use satisfaction, among other aspects, and "managers" need a way to evaluate these virtual environments that allow them to make decisions that improve their effectiveness.

The informatization of content and the need to provide agile communication and adequate management to transfer information online depends on specific technologies and to carry out the correct insertion of a web site one must understand the concept of technology of information and communication as "being much broader than just data processing, software engineering and hardware, and consideration should be given to the human, administrative and organizational aspects" [4]. For Laudon and Laudon [5] "the term Information Technology is restricted to technical aspects such as hardware and software. On the other hand, Stair [8] uses the computer-based term "Information System" to group the collection, manipulation and processing of data into information by tangible (hardware) and intangible elements (software, database, telecommunications, people and procedures) [4] concerning the information and people involved.

ICT uses technological resources to facilitate the communication and reach of information among individuals with a common goal. This technology has spread all over the world and currently plays a key role in the information and communication strategies of public and private companies. Wang [6] Gonçalves [7], this is because the digital medium allows instant access, innovating the standard of traditional services. By the vision of [4], "The strategic importance of IT was definitely incorporated in companies" (…) and "At the end of the decade, the Internet reinforced this trend, and witnessed the birth of e-business, which No doubt marks the beginning of a new era in enterprise computing." Laurindo [12], states that the success of IT management depends on two factors: efficiency and effectiveness. The efficiency, regarding the accuracy, quality and performance of equipment and software. Efficiency by function of the structural organizational part, which can influence or modify the business strategy, and by the compatibility between systems and the use of computers, making the objectives reach the needs of the organization and the final users. However, for Nielsen [8] the relationship with human aspects is also an important concept to be considered since, as he explains: "usability has assumed a much greater importance in the Internet economy than in the past."

Usability is even more important among aspects of the virtual environment than of physical environments, especially regarding the performance of a website interface. According to [8], "users today experience the usability of a site before committing to use it" (…), understanding that the web and the physical environment are different and require different approaches, where the architecture and structure of the websites should mirror the users' tasks and their visions of information space, and to define them it is necessary to understand what information, services and processes should be considered. For this purpose, he proposed the ten general principles of the user interface de-sign [8], called "heuristics", and identifies the notorious need for changes in thinking about education and communication.

Per [9] "university websites tend to follow a certain set of conventions, many of which have not changed much over the years." However, in the age of convergence, dialogues and conventions should not contain irrelevant or seldom necessary information. For [8] "The more that users' expectations are correct, the more they will feel in control of the system and the more they will like it."

Following the pace set by the information and communication society, academic websites have become essential tools for communicating between universities and their users in the virtual environment, acting as facilitators of information and dialogue with students, teachers and staff of institutions, and Principally, taking into account the needs of users away from physical installations. However, due to the various issues involved, several aspects must be analyzed so that a site can be properly representative of an institution. A university site must convey tradition, quality, brand strength and branding, credibility, excellence and innovation, without distortion and without creating misunderstandings of interpretation or communication noises that undermine this information. For Sherwin [9] "university websites tend to follow a certain set of conventions, many of which have not changed much over the years. "However, in the age of convergence, dialogues and conventions should not contain irrelevant or rarely necessary information. For Nielsen [8], "The more expectations users are up front, the more they will feel in control of the system and the more they will like it."

The websites of universities present a huge range of clients, ranging from prospective students in their courses to their alumni, through the dissemination of their research, performance in society, and other relevant information back to a highly diverse audience. Turning it into an efficient communication channel becomes complex and difficult, as it must understand each user's specific needs and goals, involving a set of distinct and dynamic information, that keeps users informed about what is happening and being in a colloquial language and of easy recognition for the internaut, following a natural and logical order [8]. These characteristics impose the need and ability to exercise a perspicacious and creative technique to communicate with users, staying within their purposes and not running away from their objectives, since the journalistic format is widely used On the web, even on sites that are not considered as such [7], and a misleading message through this communication channel could cause enormous damage to the image and brand of Universities, since the site is part of the university itself and can Be considered as an element of their identity.

This work has the objective of presenting the analysis of the first impressions that the University Websites cause to those who access them, evaluating, and identifying the reaction that would have to its design and ease of use, besides showing a methodology that makes possible an analysis That will lead these sites to increase their usability indexes. The results were extracted from a survey conducted through the internet and allowed for comparison, evaluation, and analysis from the websites of five universities: Harvard University, Massachusetts Institute of Technology, Stanford University, University of California, Los Angeles, University of Cambridge. The choice came from being among the top five in the world according to the rankings: Times Higher Education World University Rankings 2016–2017; CWUR 2016 - World University Rankings; QS World University Ranks 2016–2017; Academic

Ranking of World Universities - 2015, which provided the creation of concepts from the analyzes collected and the supporting results.

2 Evaluation and Results

The evaluation of the sites of the universities starts from the analysis found in [22] that shows that the focus should be placed on the user, and that in his Design of interface the designer must take into account variables that directly involve who will access it, such as: Expectations and needs; Respect to available time, website consistency; Ease of achievement of objectives, beyond the level of user specialization, learning ability, ability to arouse interest, user intension, among others.

A questionnaire composed of 22 (twenty-two) statements was developed to be answered through structured (closed) and unstructured (open) responses made available on the internet, so that respondents' randomness prevailed. For analysis was performed starting from the return of 100 answers and as the research was carried out in Brazil, those that required basic understanding of the English language of the respondents who did not have this knowledge were eliminated. The presentation and considerations of the results will be made omitting the name of the evaluated universities, having their identifications made by means of letters, linked in a way that did not occur their identifications.

The characterization of the participants shows that 25% have at least the degree of Ph.D., 15% of master, 5% of specialist; 85% of whom have completed a university degree and 48% are still studying. The "E" University was the most sought after to be evaluated (40%), both by those who stopped their studies in higher education and by those who have Ph.D. (24%). Figure 1 shows the choice of the Universities to be evaluated per the degree of education of the participants.

Expectations regarding site content were fully met for 67% of respondents, with only 9% responding to this item. Stratifying the research by level of education, it is identified that the highest index is found among those who only have the complete upper course (40%), followed by those who hold the Ph.D. (15%) or larger. Among those who did not meet the expectation, 44% of those who have only a higher education degree with the holders of the Ph.D. degree, as shown in Fig. 2.

The first impressions that the respondents had when visiting the sites are presented in Fig. 3. On a scale from 1 to 5, the best impression when first viewing the website visited was on behalf of the University "C", Obtaining an average evaluation of 4.3, followed very closely by University "B" with 4.29. The University of California received the lowest score, with a mean of 3.23. When the aesthetics of the site was questioned, the average of the "C" University fell to 3.9, passing the "E" University to be the best evaluated with a mean of 4.15. Again "B" University was very close, with 4.12. The evaluation of the size of the texts/letters shows that the present/used in the site of the "B" University received higher scores (average of 4.59) and those in the "D" University the lowest, with 3, 18 average. The colors used by the "C" University were the ones that pleased the evaluators. The best design, receiving an average score above all others, was that of University "B", as presented in Fig. 3. At this point it is important the method used to identify how and where the usability of a site can be

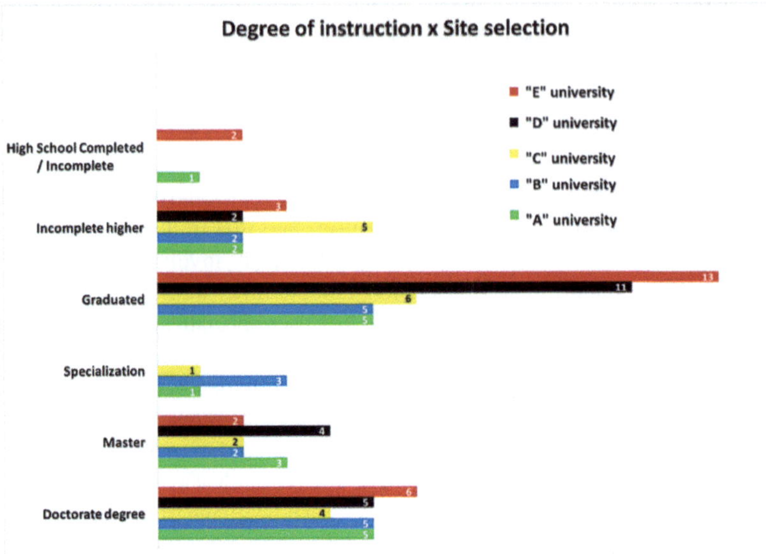

Fig. 1. Degree of schooling. x website selected

improved. From the slope of the lines in relation to the horizontal axis that join the averages obtained by the items surveyed, it is possible to distinguish the difference of values of these variables and comparatively to know which ones have more weight. Changing features would lead people who had access to the site to evaluate it differently.

Analyzing the Fig. 3, it is possible to identify that the largest angle to the x-axis of the graph occurs between the averages obtained by the "B" University for the items "aesthetically attractive" (4.12) and text/letter size "(4,59). An increase in the average score obtained in "aesthetically attractive" will reflect directly on the first impression people will have when they access the site, improving their evaluation. As this site received a high average in the "letter/text" question and the average of the "suitable/ pleasant colors" item is above that achieved by "aesthetically attractive", there is a need to rethink its design and information distribution Contained therein. A rearrangement of its distribution or colors could also lead to better notes, once a sharp slope is identified. Ideally, the angle between two subsequent lines should always be as close to zero as possible, with the highest values possible. This same analysis can be applied to other sites, such as University "A". This site was rated a Regular, with ratings of its aesthetics and colors with averages below the 4.0 mark. Changing these items would lead to the site getting a better final rating. This method of analysis is quite simple and powerful and with the choice of the appropriate items for evaluation, it is possible to identify where to change the website to increase its usability index.

The degree of difficulty in finding information within the sites is presented in Fig. 4, with the greater and nearer to note 5 being the value attributed, the less difficulty and the easier the information requested. This evaluation portrays the users' perceptions regarding the time taken to arrive at the desired information, as well as their

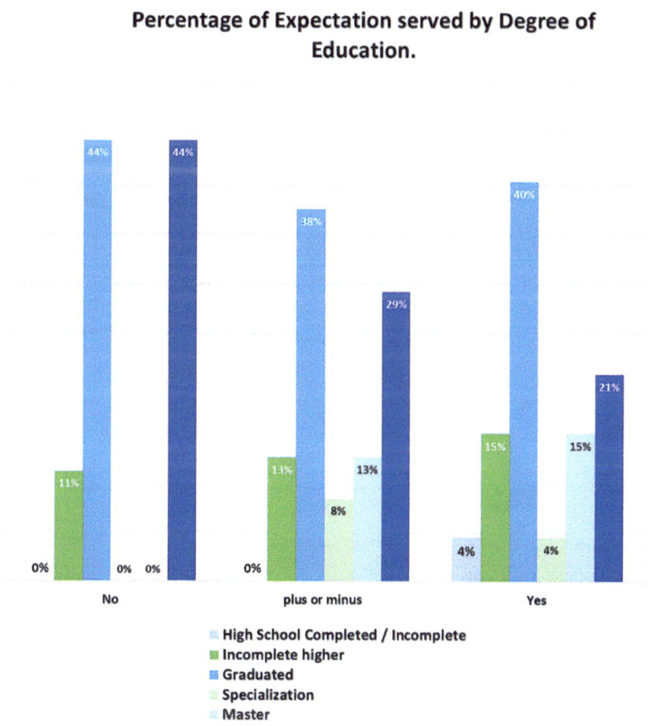

Fig. 2. Expectations results about site x degree of education

interaction with the available interface. In Fig. 4, where the averages obtained are represented, it is also portrayed the level of user experience regarding access to university websites.

The evaluation showed that, in general, there was no difficulty in finding most of the requested information, as well as the design of its interfaces facilitated navigation. Exception was due to the difficulty in finding the information related to teachers, where the averages were between 2.5 and 3.3, with note 5 being related to the easiest level.

The frequency of access to university sites shows that even those who have the greatest experience in navigating university websites (average 3.61) and who chose "C" University, attributed a degree of excessive difficulty in the search for Information about teachers (2,5) of this University. This information shows that communication on this topic should be improved.

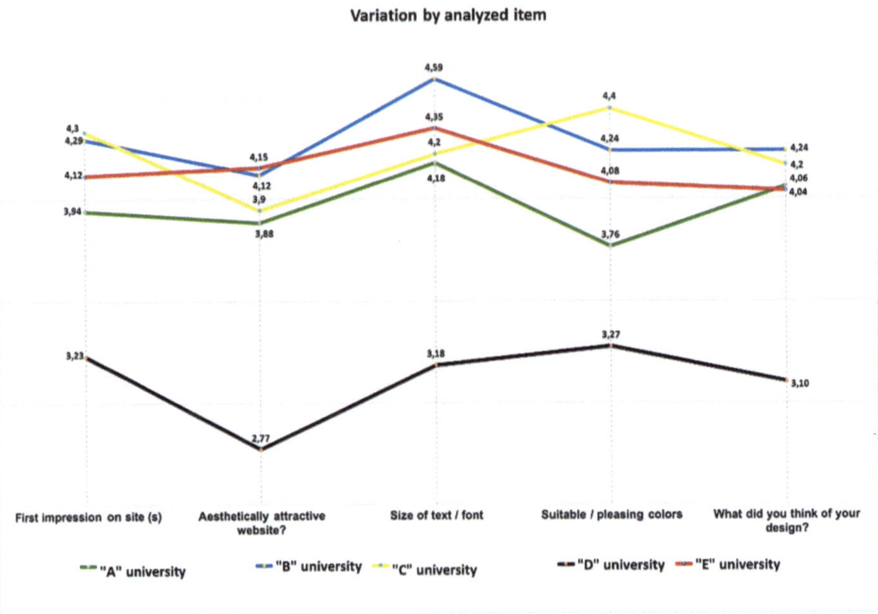

Fig. 3. Impressions about the Universities websites

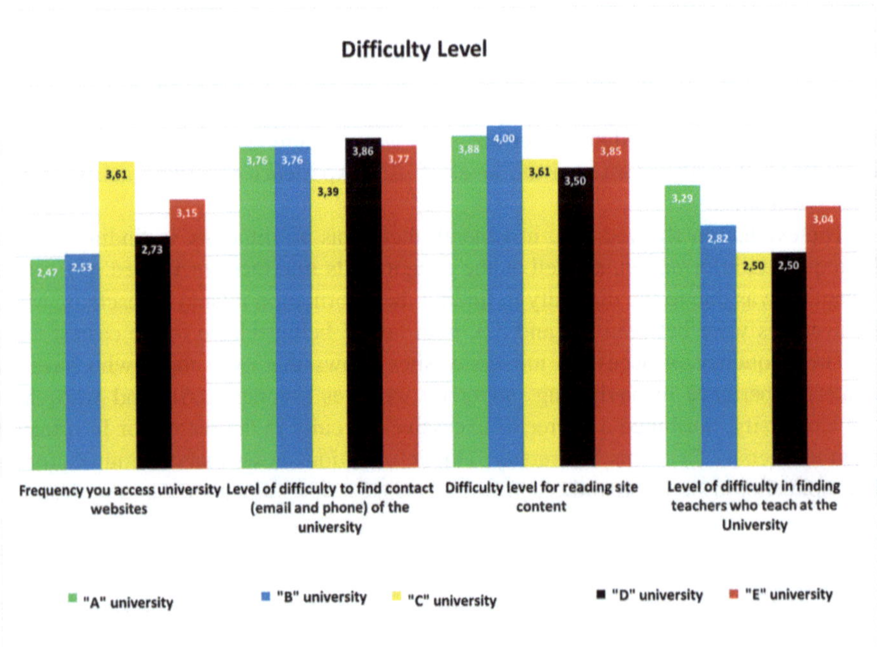

Fig. 4. Impressions about the Universities websites

3 Conclusion

The results of the survey when tabulated considering the average of opinions show that there is a heterogeneity for the first impression provided by the design caused used on the university websites, even among those belonging to the considered best universities, per published rankings. In an evaluation in which the maximum score for these items, which relate color, text and design was 5, there was site in which the average of their ratings did not reach 3.3, while others had most of their scores above 4.0, even getting an average above 4.5 for a specific item. The degree of difficulty in navigating and obtaining information was another point analyzed, and none of the sites obtained average scores above 4.0. In this item, there was an evaluation in which the average did not exceed 2.5 showing that there is a great margin, still, for it to be improved.

The method of analysis presented using the slope of two subsequent lines joining the notes related to the items surveyed was both simple and powerful, making it possible to identify where to intervene in order to improve the usability of a site.

The factors presented here are important and must be constantly monitored as the strength of the communication channel provided by the websites becomes bigger, making them the first contact with the Institution and therefore part of his brand and personality. A site that meets expectations and is quickly finding information without the user spending more time than necessary and helping you achieve your goals quickly will make a difference in the final decision that you will make. That is why interface designers should intensify the focus on the user, placing it as the central reason for the existence of the site, so that its interaction with the interface provides satisfaction when browsing, searching, and finding the information it needs, and who accesses it for the first few times, the ease of assimilating how to proceed and achieve its goals. That the user's explicit and implicit intentions are met and that an institutional message is presented that represents the greatness and values of the organization.

References

1. de Souza, C.S., Prates, R.O., Barbosa, S.D.J.: A method for evaluating software communicability. In: Anais do II Workshop sobre Fatores Humanos em Sistemas Computacionais (IHC 1999), Campinas, Artigo 28 (1999)
2. Prates, R.O., Barbosa, S.D.J., de Souza, C.S.: A case study for evaluating interface design through communicability. In: Proceedings of the International Conference on Designing Interactive Systems, DIS2000. ACM Press, New York, 308–317 (2000)
3. Lopes, L.A., Melão, N.F.: O Desempenho De Websites Organizacionais: Uma Proposta de Modelo Avaliador. Gestão e Desenvolvimento **21**(2013), 57–81 (2013)
4. Borges, T.N., Parisi, C., Gil, A.L.: O Controller como Gestor da Tecnologia da Informação - Realidade ou Ficção? Revista de Administração Contemporânea **9**(4), 119–140 (2005). http://dx.doi.org/10.1590/S1415-65552005000400007
5. Laudon, K.C., Laudon, J.P.: Gerenciamento de Sistemas de Informação, 3rd edn., p. 7, 170, 228, 344. LTC, Rio de Janeiro (2001)
6. Wang, C.B.:. O novo papel do executivo de informática, 1st edn., p. 58. Makron Books, São Paulo (1995)

7. Gonçalves, S.V., dos Santos, N., Teixeira, C.S., Macedo, M.: Formato, notícia e curadoria digital em sites de parques brasileiros
8. Nielsen, J.: Projetando Websites: Designing Web Usability. Elsevier Editora, Rio de Janeiro (2000)
9. Sherwin, K.: Breaking Web Design Conventions = Breaking the User Experience (2014). https://www.nngroup.com/articles/breaking-web-conventions/

The Style of Imagery: A Color Study of Landscape Paintings

Lung-Wen Kuo[1,2(✉)] and Chih-Chun Lai[3]

[1] Department of Multimedia Design,
Takming University of Science and Technology, Taipei 114, Taiwan
Kuo@ttu.edu.tw
[2] Graduate Institute of Design Science, Tatung University, Taipei 104, Taiwan
[3] Department of Industrial Design, Tatung University, Taipei, Taiwan
lai@ttu.edu.tw

Abstract. The use of color is a crucial element in painting, since any particular color would be affected by its surrounding colors in order to produce a different visual hue. This study focused on the study of landscape paintings and color, based on the study of art history, and investigates approaches via analysis of colors adopted by impressionist painters in western paintings. This study explores the elements of hues, brightness, saturation, complementary colors, and cold/warm tones in contrasting colors, as well as corroborating these elements in actual artwork. Landscape paintings primarily depicts local geographical and cultural features, while colors convey emotions and feelings. A work of depth exhibits a strong sense of emotions, integrates scenery and features to produce a theme with space for imagination and innovation.

Keywords: Color · Research · Painting

1 Introduction

The innovative use of colors in painting is one of the important elements of the picture characteristics, for example when configuring colors, small features will be affect by the color selection, this is called the color contrast phenomenon. Color contrast affects hue, lightness, saturation, fill, warmth and so cold perception, this is a method of expression and allows the painting to be presented with a screen color and reference color using ancient genre painter. The method is the result of the necessary in-depth knowledge and research, motivated by the author research.

In modern renowned color guru Johns, Eden (Johannes Itten, 1888–1967) book "Color Theory", "Eden twelve hue ring" is easy to understand with the color basis. Its design features are the three primary colors foundation hue. Among twelve color grid are the same, and six complementary color pairs, which are located in opposite ends of the diameter; the development of a second color wheel. Twelve hue ring is composed of primary colors (primary colors) (red, yellow, blue), a secondary color (secondary colors) (mixing any two of the primary income of colors) and three color (tertiary colors) (with two adjacent colors secondary colors mixed colors) combination.

A.G. Ho (ed.), *Advances in Communication of Design*, Advances in Intelligent Systems and Computing 609, DOI 10.1007/978-3-319-60477-0_8

2 Method

2.1 The Relationship of Color and Light

Elements of color and light define the main characteristics of paintings. In impressionism, the "light" defines real objects, and is considered one of the basic principles of painting techniques with color decomposition. The painter considers the object itself unimportant, but by changing the light around the object, and the shadow beneath the subtle colors, the painting theme is decided. Color development was based on a scientific theory (Rainbow analysis generated by sunlight) "spectral colors" painting. Impressionist painter Monet (Claude Monet, 1840–1926) indicated that even the darkest shadows on the screen should also be colored, and composed of light and color at different levels posed almost imperceptible, but color brightness and saturation can be low, therefore Monnet often described the same subject in a different lights constantly by changing the appearance in the painting in order not to see the object itself, but the process of change observed with the light.

Monet often painted the same landscape at different times, paying attention to the appearance of constantly changing color under different light and air, showing the effects of time-varying lighting is presented, the theme of water lily series painted forty-eight works. Church of Rouen (Rouen Cathedral) have thirty paintings, the painter wanted to illustrate the color change of light and shadow projected onto the front wall of the church, that is, to show the relationship of color and light and church buildings.

2.2 Color System

Everyday themes usually divided by Yiba or "color" and "black and white". No color: refers to color, including white, black and a brightness change between all levels of gray. No color should be bright or dim by the degree to distinguish. Chinese ink painting can be said that no successful examples of the use of color, Five Dynasties of Li Cheng, specializes in painting landscapes, and the landscape of the performance of the content and presentation skills, to the development of the depth of his landscape paintings not only show a change in the weather in the mountains, also emphasized the seasonal climate characteristics, the "reading stone monument nest diagram", the whole works in pen and ink (no color) is completed, the whole picture extending into the distance, the farther the more light, shrouded in fog, adds a bit Mystery.

Early Cubist Picasso (Pablo Picasso, 1881–1973) and Braque (Georges Braque, 1882–1963) used to paint with gray tone, the main emphasis of the characteristic composition analysis, to be decomposed to depict objects, then the object while the image of the respective faces painted on canvas, the work looks like a pile of blocks, noodles piled together haphazardly constructed; the color is closer to a single hue. Picasso's "Guernica" (Guernica), the use of black, white and gray tones to present, making the whole picture of the screen looks dark unknown, shape and color (no color) chiaroscuro, makes color-forming The atmosphere is full of sorrow, a witness to the tragedy of the times, painting the front of the woman's face and the side face binding, it is Cubist movement of the viewpoint of the painting.

Color: refers to all colors other than the achromatic color, regardless of the level of saturation are included, not only according to different colors respectively, as well as light and dark shades and are therefore very much a color number. A color can be mixed unlimited color variations (Fig. 1).

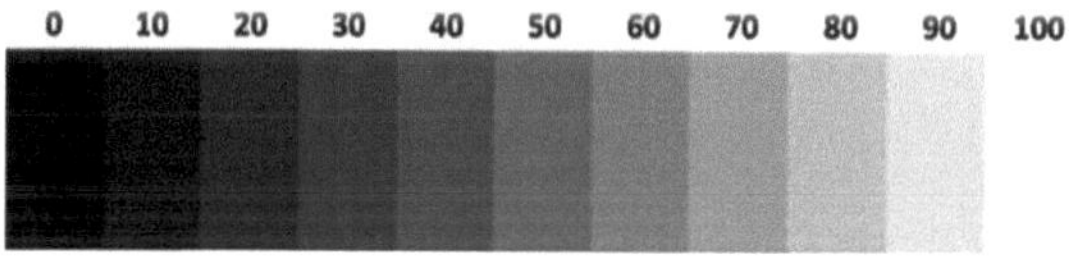

Fig. 1. Achromatic color gradation

2.3 Color Properties

Any color has three attributes, namely, the three elements of color are: Hue (Hue), lightness (Value), saturation (Chroma), any color must have these three basic conditions. Hue refers to the color appearance, color difference name, regardless of any color, lightness and saturation of the three properties reflect changes, visual senses are formed in the changing colors, such as "Yellow". Hue basic colors red, orange, yellow, green, blue, cyan, purple and other colors for the levels, in order to expand the color changes, as the basis of the hue and form a three-dimensional gradient colors, colors have a three-dimensional properties.

Saturation refers to the color saturation (saturation), solid phase for the highest chroma colors, the highest color saturation that is known as a solid color. When hue and white is introduced they will reduce color saturation, solid color plus white light colors are formed, known as the "bright color." The more light, the brighter the color, the more visual sense light, cool, lively. When mixed with solid black color darkens, saturation is lower, known as the "dark", the darker the color the deeper, more visual feel heavy, dark, athletic. While adding black and white saturation will be reduced, the lower the degree of gray plus more colorful, plus gray color formed after known as "intermediate color." Change saturation.

In "hydrangea", the theme is hydrangea flowers and background to distinguish brightness and contrast, color saturation with halftone, light purple petals, with flowers and three-dimensional texture, background faintly visible to other flowers and green leaves, and the flowers and the background even as a whole, showed purple tones of beauty.

This four-color three-dimensional work of Mansell, represents the "alienation" brown hue, lightness and saturation dark and cloudy. "Lily" is yellow-green hue, lightness and saturation relatively moderate. "Hydrangea" blue-purple hue, lightness bright flowers, chroma more neutral tones. "Tung Blossom" green hue, lightness bright flowers, chroma neutral tones as shown in Fig. 2.

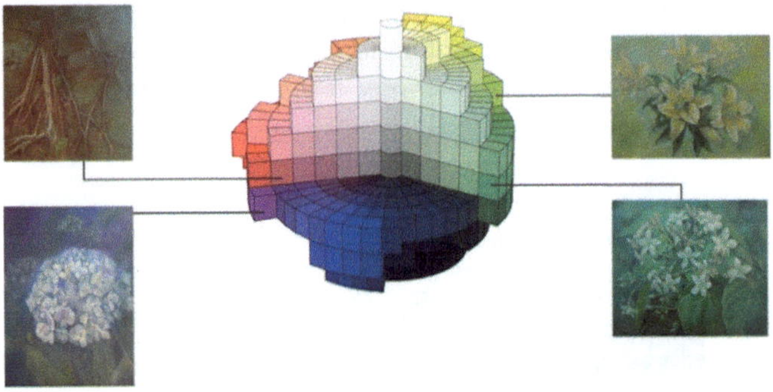

Fig. 2. Works with Mansell color perspective

3 Results

When painting, color harmony helps developing the perception and performance of the whole painting representation, and helps achieve the desired color on the screen beauty, the harmony of color is reflected with orderly color combinations, with the intention of mood reflection for deployment of a engender pleasure or special surprise sensory stimulation. Use of color hue, brightness, and color change saturation of the area occupied by the position to adjust the color of unity and variety, the color of harmony can be achieved, increasing the richness and beauty of the screen.

Tone is not holistic perception of the painting, in the painting, it's like the use of blue, green tones, warm or cool colors, etc. If paintings not a uniform color, the more it will appear disorganized. There are no shades of color and the color between different hue, lightness and saturation with interaction from the overall relationship, with bright colors, light colors, light colors, dark colors, dark shades, turbid color tone. In Hue points, pure colors, cool colors, warm colors and so on.

The Author creates "wild white together" in the use of the neutral shades of green for the tracking, background, leaves, stems, flowers depicted below, eighty percent of the entire painting color, are using different color classes times green to show, to create a blending colors on. While Lily was originally white, but the performance of three-dimensional petals shadow detail depicting flowers are added to green elements into account, so that the screen is more unified feeling tone, to make the picture more variability, and a similar background Impressionism Pointillism technique new law will be a variety of blue, orange, and green dots of different colors, distributed on the screen, the screen adds dynamic and lively, close to the flowers and leaves of space, places a lower brightness color, make flowers, foliage and background grated sense of space, highlight the theme of integrity, but also to coordinate a more complete picture.

Distinguishing colors, usually reflects sensual visual differences, blue, gray, white, blue and other colors will feel the sky, water, ocean, snow, cold called cool colors or tones; red, orange, yellow and other colors will feel sun, fire, warm, it is called warm; low saturation, low brightness, dark colors, dark tones, shades and other cloud is relatively calm tones, and will feel static, steady and deep.

In the painting "frog swimming" dark colors and shades of dark green and deep purple cloud based colors were used, the dark green water with chiaroscuro rendering space texture water, water ripples also with shades of deep green, reflecting presenting dynamic and light waves, the water above the light changes to render the feeling of reflection and object to the saturation of green, purple color elements, interactive use, so that richer color screen and blending. Warm brown frog itself, in order to reflect more explicit stance color, the higher the water saturation with brown rendered underwater is a combination of color types, saturation helps lower brightness in the brown, which makes a transparent surface feeling. Water Lotus Ming with high yellow-green, dark shadows and objects with the surface of the dark color, the performance of the overall texture, so that the different elements are also linked to the intersection, the larger the surface because of the background of the area occupied, there will be a whole painting feel unified tone.

In the "Water Lilies" in order to utilize lighter colors and high chroma, a yellow green blue-green color was used, water lily leaf distributed below the screen deliberately creating a unity of tone picture. Lily used to change the brightness of the green color gradation and deep blue shadows to lighting effects and three-dimensional sense manifested on the water in a more vivid blue-green branches in the leaves around, farther from the center of the frame color, the lower the brightness, saturation the water leaves the lake obviously separates produce layered, and bring out the importance of the theme of water lilies, but also to make the picture more colorful and varied. Distant places lower saturation of water lily purple, green, blue and other simple colors, color, explain the sense of space and distance on the lake front, and two flower decorated with warm lavender lotus flowers, timely appear, so the picture will not be too monotonous, but with more vitality perception (Fig. 3).

Fig. 3. Cool and warm tones corresponding to fig.

4 Discussion

4.1 Color Contrast

During simultaneous color comparison, for two colors, two colors each will affect the other to produce different contrasts, where color will be affected by other surrounding colors to produce the visual illusion. When two different brightness of colors juxtaposed, it makes clear color is more bright, and dark colors darker. When saturation contrast, color saturation and different set generated by contrast, low-saturation color configuration will appear more vivid, and high saturation and make the colors appear more turbid set. Complementary color is relative colors in the color ring when the two are in a relative position, contrast two complementary color when juxtaposed will produce one of them if only to see the complementary color, will form an appropriate complementary color illusion in the retina of the eye phenomenon. In the Fauves (Fauvism) Matisse (Henri Matisse, 1869–1954) for the "Music" in order to color flat-coating method, the use of low brightness dark shades painted sky blue, to bring out the chiaroscuro of the human body; bright green painted surface, high performance chroma bright red body color, between the three colors of light and solid, resulting in a harmonious visual effect.

In the "Lily" and "Morning Glory", painting if Fig. 4, the use of the concept of contrasting color to blue and orange color complementary effects after using orange and blue showing more vivid and bright, contrasting colors using the color way to make Color works more convex, and works will change for the better.

Fig. 4. Works with morning glory lily's correspond to each other using the color circle in comparison

4.2 Mixed Colors

Painting with blended colors can affect the presentation of any painting. For example if one compares spectral colors impressionist painting with a real canvas shade mixed with pigments, the results shows different styles. When adding two or more shades, the effect of added brightness will highlight the effect around shaded objects, but the same amount of shade mixed with primary colors will overlap and form white coated pigments to have various results. Color mixing in oil painting with subtraction technique is achieved when the number of pigment are mixed, also with the increase in the hue, the painter produce lighter effects, while chroma reduction helps in cloudy gray coloring, blending with the increase in the number of color provider darker shades.

References

1. Li, C.: Outline of Western Art History. Lion, Taipei (1979)
2. Xiong, X., Travel, P., Pi, N.S.: China's circulation coins legibility of the assessment study. Technol. J. **7**(1), 103–109 (1991)
3. Lin, S.: Color Science. San Min, Taipei (1993)
4. Hou, L.: Evolution of the Western Landscape Painting Studies. China Bookstore, Taipei (1978)
5. Zhang, X.: Landscape Painting - Western Landscape Masterpieces Appreciation. Lion, Taipei (2002)
6. Dai, M.: Modern Chromatics. Full Hua, Taipei (1995)
7. Fairview Press. Western painting in 2000, Taipei (1999)
8. Ping, P.: Secret Monnet. Art Books, Taipei (1996)
9. Hsun, C.: Introduction to the Arts. Donghua Bookstore, Taipei (1995)
10. Liu, Z.: Impressionist Paintings. Art Books, Taipei (1998)
11. Hideaki Europe: Application of Color Science. Lion, Taipei (2002)
12. Zhao, H.: Art Appreciation. San Min, Taipei (2002)
13. Xi, Z.: Color Science. Five South, Taipei (2007)
14. Zhang, X.: Tour the History of Western Art. Lion, Taipei (2001)
15. Wei, C.: Practical Chromatics. Holley Books, Taipei (2006)

The Effectiveness of Message Sidedness on Trust in Check-in Advertising

Chi-Lun Liu[1(✉)], Jo-Han Ku[1], and Hsieh-Hong Huang[2]

[1] Department of Multimedia and Mobile Commerce, School of Informatics, Kainan University, 1, Kainan Road, Taoyuan City 33857, Taiwan
tonyliu@mail.knu.edu.tw, a0918822380@gmail.com

[2] Department of Information Science and Management Systems, National Taitung University, 684, Sec. 1, Chunghua Road, Taitung 95002, Taiwan
kory@nttu.edu.tw

Abstract. Due to the popularity of social websites and smart phones, check-in services are regarded by companies as a means of consumer-generated advertising. This study explores the influences of check-in advertising on trusting beliefs for check-in content. This study is based on the media richness theory and the attribution theory to propose the research model. The experiment method is used to evaluate the proposed research model. The results show that, in trusting beliefs for check-in content, two-sided content is better than the content without text. In the research contribution, this research explores the applicability of media richness theory and attribution theory in check-in advertising of social media. Companies can apply the research findings to facilitate effective communication in check-in advertising.

Keywords: Sidedness · Social media · Check-in service · Trust · Mobile marketing

1 Introduction

Social media is emergent as a powerful communication tool has made it possible for a person to communicate other people about products and companies [1]. Many social media users, such as Facebook and Instagram users, like to use check-in services to post messages and photos about their lives in specific locations. Check-in service is a location-based and one-to-many model that allows users to create and name venues to manually broadcast their locations to their friends and others in social media [2]. Social media users post contents in restaurants voluntarily. And many physical restaurants encourage consumers generate check-in contents to share their comments on their restaurants for advertising purpose.

The content posted by check-in service users in a restaurant often have pros and cons about the restaurant. A one-sided advertisement only describes positive attributes of a product. And a two-sided advertisement describes both positive and negative attributes of a product. The prior sidedness studies in advertising communications [3–5] investigated on the effects of pros and cons about products and services in the offline

A.G. Ho (ed.), *Advances in Communication of Design*, Advances in Intelligent Systems and Computing 609, DOI 10.1007/978-3-319-60477-0_9

environment. Online advertising and social media advertising are mainstream advertising channels in recent years. Therefore exploring the effectiveness of sidedness is an interesting research topic in social media application context. Some emergent studies explore advertising effectiveness in social media [6, 7]. Many studies focus on advertising reliability and trust [8].

Media richness theory describes that there is a relationship between the richness of communication media and the communication ability [9]. Attribution theory reveals that a causal explanation process in which an individual assigns causes to events [10]. In tradition, advertisements always describe positive attributes of a product because companies have to sale their products to earn benefits. Negative attributes of a product may be hidden in advertisements. According to attribution theory, a two-sided advertisement may more credible than an one-sided advertisement because a two-sided advertisement which have both positive and negative attributes provides a fair analysis about a product.

The research objective of this study aims to manipulate sidedness in the check-in based advertisement and to explore how sidedness influences trusting beliefs about the check-in content. Media richness theory and attribution theory are used as the theoretical foundation of this study. An experiment is conducted to provide empirical evidences about the relationships between sidedness and trusting beliefs.

2 Three Message Designs in Social Media Check-in Contents

This study uses the popular Facebook app, which provides check-in services on both android and iphone platform in the experiment. Users can utilize the Facebook app check-in service to share their location in their user-generated contents. The Facebook app users can alternatively post text message and photos in their check-in contents.

Fig. 1. Check-in content design C-1: no text.

Fig. 2. Check-in content design C-2: one-sided text.

Fig. 3. Check-in content design C-3: two-sided text.

Three text message designs for photo description are usually appears on user-generated check-in content in the practices: no text, on-sided text, and two-sided text. This study uses the Facebook app and the mobile device, which is Samsung Note3, to manipulate three check-in content designs: C-1, C-2, and C-3. A user only posts a photo in check-in content C-1 (Fig. 1). A user posts an one-sided text message which describes the advantage of the shabu shabu restaurant and a photo in check-in content C-2 (Fig. 2). And a user posts a two-sided text message which describes the advantage and disadvantage of the shabu shabu restaurant and a photo in check-in content C-3 (Fig. 3).

3 Research Hypothesis

Figure 1 depicts a proposed research model. The independent variable is check-in content. And the dependent variable is the effectiveness in content trusting beliefs. The proposed three hypotheses (H1, H2, and H3) in Fig. 4 are introduced as follows.

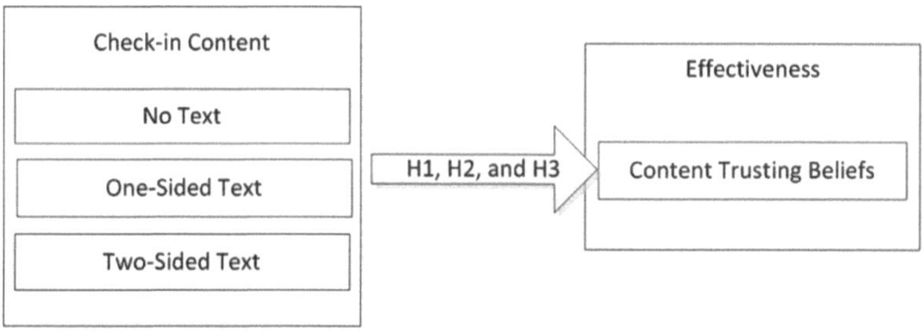

Fig. 4. Research model

Message sidedness is the style of argument in message content. Message sidedness includes one-sided and two-sided messages. The one-sided message presents the agreeable argument in favor of a particular position, i.e. positive information. And the two-sided message presents the specific argument but also the opposing arguments [11], i.e. both positive and negative information. The effects of message sidedness can be explained by attribution theory. *Attribution theory* focuses on receiver's interpretation of a sender's motives for communicating such information [12].

Media richness theory indicates that media differing in cues variety offers different communication ability [13]. Text and photo are two cues. Higher levels of media richness on web site selling apparel have positive effect on customer trust in company's web site [14]. Trusting belief is a perceiveness on trust. With *trusting beliefs*, a receiver perceives that sender generates promised and reliable content, even though doing so could involve risks [15]. Hence the hypothesis 1–2 are proposed as follows.

- *Hypothesis 1 (H1):* Respondents exposed to two-sided check-in content have higher level of trusting beliefs than those exposed to no text check-in content.
- *Hypothesis 2 (H2):* Respondents exposed to one-sided check-in content have higher level of trusting beliefs than those exposed to no text check-in content.

Comparing to one-sided advertisement including only positive information, two-sided advertisement includes positive and negative information enhances the source credibility because the attribution theory indicates that the sender who post two-sided advertisement is more honest than the sender who post one-sided advertisement [8]. Therefore the hypothesis 3 is proposed as follows.

- *Hypothesis 3 (H3):* Respondents exposed to two-sided check-in content have higher level of trusting beliefs than those exposed to one-sided check-in content.

4 Research Method

This study manipulates media richness and sidedness in the experiment setting. Table 1 summarizes two levels of media richness and three types of sidedness in check-in contents. Subjects were recruited in a university and several companies are randomly assigned in three groups to browse check-in content C-1, C-2, and C-3 on Facebook. The subjects were Taiwanese.

Table 1. Media richness and sidedness in check-in content design

Check-in content design	Media richness	Sidedness	Features
C-1	One cue	None	Text: There is no text Photo: The photo shows the dishes, the environment and the consumer in a restaurant Map: The geographic location in the map shows when a user clicks the location name
C-2	Two cues	One-sided message	Text: The message includes a positive attribute of the dishes Photo: The photo in C-2 is the same as the photo in C-1 Map: The geographic location in the map shows when a user clicks the location name
C-3	Two cues	Two-sided message	Text: The message includes a positive attribute and a negative attribute of the dishes Photo: The photo in C-3 is the same as the photo in C-1 Map: The geographic location in the map shows when a user clicks the location name

After the subjects browse a particular check-in advertisement, the subjects fill in the questionnaire to measure their trust beliefs about the check-in advertisement. The items in the questionnaire for measuring trusting beliefs is developed based on the electronic commerce study [15].

5 Experimental Result

The experimental result is twofold. The first part is descriptive statistic data for reporting the subjects' background. The second part is the hypotheses testing result conducted by ANOVA. The two parts are introduced as follows.

5.1 Descriptive Statistics

The descriptive statistics reports the user experiences for three groups of social cues designs (C-1, C-2, and C-3) about trusting beliefs. The sample size is 165. The statistic data is shown and discussed as follows.

Table 2 shows the subject demographics. In the experiment, male subjects are about 61% and female subjects are about 39%. Most of subjects are 24–28 and 34–43 years old. All subjects have usage experiences on Facebook. 89% subjects used the Facebook check-in service. Most of subjects are office Staves.

Table 2. Subject demographics

Gender	
Male	60.6%
Female	39.4%
Age group	
Under 18	2.6%
19–23	23.9%
24–28	31.0%
29–33	11.6%
34–43	28.4%
44–53	1.3%
Over 54	1.3%
Occupation	
Student	30.5%
Office staff	69.5%
Have Facebook usage experience	100%
Have check-in experience	89.0%

Table 3. Agreed percentage on trusting beliefs

Media richness	One cue	Two cues	
Sidedness	None	One-sided	Two-sided
	C-1	C-2	C-3
I believe that the check-in content which describes the characteristics of the restaurant	52.3%	65.3%	76.5%
I believe that the check-in content is helpful in choosing a good restaurant	49.2%	63.3%	67.2%
The check-in content is trustworthy	46.2%	63.3%	64.7%
I believe that the check-in content does not introduce a bad restaurant	52.3%	59.2%	64.7%
	50.0%	62.8%	68.3%

Table 3 shows that higher social cues richness results in more subjects who have trusting beliefs on the content in average agreed percentage. From the message sidedness standpoint, 68.3% subjects who read the two-sided message agreed that they believe the content, 62.8% subjects who read the one-sided message agreed that they believe the content, and only 50.0% subjects who read the photo without message agreed that they believe the content. Therefore the two-sided message is better than one-sided message and none message in trusting beliefs.

5.2 Hypotheses Testing Result

This study uses one-way ANOVA to find the differences about check-in content effectiveness among three groups. The reliability of the trusting beliefs dimension is acceptable because the value of cronbach's alpha is 0.877. The Tukey-Kramer method is used in multiple comparisons. The preliminary results depicted in Table 4 shows that the trusting beliefs in the two-sided text (C-3) group is significantly higher than the trusting beliefs in the no text (C-1) group. Therefore hypothesis H1 is supported.

Table 4. Comparisons for group differences on trusting beliefs

Check-in content	Average	Standard deviation	Trusting beliefs	
			Mean difference (I-J)	P value
No Text (I)	14.0308	2.57372	−0.37739	0.718
One-sided text (J)	14.4082	2.65297		
One-sided text (I)	14.4082	2.65297	−0.72909	0.334
Two-sided text (J)	15.1373	2.48209		
No text (I)	14.0308	2.57372	−1.10649	0.058*
Two-sided text (J)	15.1373	2.48209		

*indicates $p < 0.1$.

According to media richness theory, the one-sided text check-in content including two cues has higher media richness than the no text check-in content. The average of trusting beliefs in the one-sided text (C-2) group (14.4082) is slightly higher than the average of trusting beliefs in the no text (C-1) group (14.0308) in Table 4. Table 3 also shows that 62.8% subjects agree that the one-sided text check-in content is trustworthy and 50.0% subjects agree that the no text check-in content is trustworthy. Although the p value (0.718) indicates that hypothesis H2 is not supported by the ANOVA test, the descriptive statistics data shows that more subjects believe in the one-sided text check-in content than the no text check-in content.

According to attribution theory, the two-sided text check-in content provides more fair analysis than one-sided text check-in content. Table 4 shows that the average of trusting beliefs in the two-sided text (C-3) group (15.1373) is slightly higher than the average of trusting beliefs in the one-sided text (C-2) group (14.4082). Table 3 also shows that 68.3% subjects agree that the two-sided text check-in content is trustworthy and 62.8% subjects agree that the one-sided text check-in content is trustworthy. The descriptive statistics data shows that more subjects believe in the two-sided text check-in content than the one-sided text check-in content. The ANOVA test does not have sufficient evidences to supported hypothesis H3.

6 Conclusion

In theoretical contribution, this study put the concept of sidedness into the social media's check-in service as a promotion channel in the proposed research model. The proposed theoretical research model is evaluated preliminarily in the experiment. For managerial implications, if the senders of check-in content want to increase their source credibility, they can apply two-sided message and avoid no text description in their check-in contents. However, the companies, such as restaurants, can alert the negative information about products and services in the two-sided check-in contents.

References

1. Mangold, W.G., Faulds, D.J.: Social media: the new hybrid element of the promotion mix. Bus. Horiz. **52**(4), 357–365 (2009)
2. Cramer, H., Rost, M., Holmquist, L.E.: Performing a check-in: emerging practices, norms and 'conflicts' in location-sharing using foursquare. In: 13th International Conference on Human Computer Interaction with Mobile Devices and Services, pp. 57–66. ACM, New York (2011)
3. Kamins, M.A.: Celebrity and noncelebrity advertising in a two-sided context. J. Advert. Res. **29**(3), 34–42 (1989)
4. Smith, R.E., Hunt, S.D.: Attributional processes and effects in promotional situations. J. Consum. Res. **5**(3), 149–158 (1978)
5. Florack, A., Ineichen, S., Bieri, R.: The impact of regulatory focus on the effects of two-sided advertising. Soc. Cogn. **27**(1), 37–56 (2009)
6. Liu, C.-L.: The impact of social cues and effectiveness in check-in advertising. Kybernetes **43**(7), 984–1002 (2014)
7. Liu, C.-L., Huang, H.-H.: A preliminary study on social cues design in mobile check-in based advertisement. In: Nah, F.F.-H. (ed.) HCIB/HCII 2014. LNCS, vol. 8527, pp. 744–753. Springer, Heidelberg (2014)
8. Eisend, M.: Two-sided advertising: an meta-analysis. Int. J. Res. Mark. **23**(2), 187–198 (2006)
9. Daft, R.L., Lengel, R.H., Trevino, L.K.: Message equivocality, media selection, and manager performance: implications for information systems. MIS Q. **11**(3), 355–366 (1987)
10. Crowley, A.E., Hoyer, W.D.: An integrative framework for understanding two-sided persuasion. J. Consum. Res. **20**(4), 561–574 (1994)
11. Allen, M.: Meta-analysis comparing the persuasiveness of one-sided and two-sided messages. West. J. Speech Commun. **55**(4), 390–404 (1991)
12. Laczniak, R.N., DeCarlo, T.E., Ramaswami, S.N.: Consumers responses to negative word-of-mouth communication: an attribution theory perspective. J. Consum. Psychol. **11** (1), 57–73 (2001)

13. Dennis, A.R., Kinney, S.T.: Testing media richness theory in the new media: the effects of cues, feedbacks, and task equivocality. Inf. Syst. Res. **9**(3), 256–274 (1998)
14. Hassanein, K., Head, M.: The impact of infusing social presence in the web interface: an investigation across product types. Int. J. Electron. Comm. **10**(2), 31–55 (2005)
15. Sia, C.L., Lim, K.H., Leung, K., Lee, M.K.O., Huang, W.W., Benbasat, I.: Web strategies to promote internet shopping: is cultural-customization needed? MIS Q. **33**(3), 491–512 (2009)

Towards a Generalization of New Media

Edward C.K. Hung(✉)

Department of Journalism and Communication,
Faculty of Arts, Hong Kong Shue Yan University,
10 Wai Tsui Crescent, Braemar Hill Road, North Point, Hong Kong
ckhung@hksyu.edu

Abstract. With the spate of new media for different purposes, how should we study them collectively so that we can have a better understanding of them to develop their next generation? In this paper, we generalize new media from a designer's perspective and adopt digital artifacts from new media including digital video, digital music, and websites to represent new media. We find out that these digital artifacts possess many characteristics of cultural products; therefore, it is possible to use the creative work ontology, which has a special focus on cultural products, as a starting point to generalize new media. We conclude that new media have rules, structures, features, and biases for the generalization.

Keywords: Digital artifacts · Generalization · New media design

1 Introduction

Nowadays, new media is part of our lives. Its surge happened when smart phones became very affordable. Whatsapp, WeChat, Facebook, Instagram, Twitter, etc. are some of the new media names we always mention. Academically, how should we study new media collectively? How should we dissect them into different parts to prepare for a better design?

In fact, a generalization of new media can be meant for a myriad of purposes. In this paper, we limit our discussions to new media design. Herein, new media design is about designing a new media for entertainment, information exchange, and communication enhancement. The information carried by a new media enables a user to access its characteristics and grasp the way it facilitates communications. An example is a video on YouTube. Once a user has uploaded his video to YouTube, he can use the functions in YouTube to manipulate his video and find out the number of visitors who view the video. Referring to the studies by Jung and Stolterman [1] and Brummett and Duncan [2], it is possible to claim that the digital artifacts from new media such as digital videos, digital music, and websites are eligible to represent new media. We will further elaborate on this representation in the coming sections. Furthermore, in view of the study by Hung and Choy [3] about the nature of cultural products and their invention of the creative work ontology as regards the creation of cultural products, we find that the digital artifacts from new media possess many characteristics of cultural products. Hence, it is possible to refer to the creative work ontology to explore the

A.G. Ho (ed.), *Advances in Communication of Design*, Advances in Intelligent Systems and Computing 609, DOI 10.1007/978-3-319-60477-0_10

generalization of new media. Under these circumstances, our research direction is how to generalize new media for new media design when new media can be represented by their digital artifacts that carry the characteristics of cultural products. It is expected that through this research we can define a new starting point for the generalization of new media – the digital artifacts of new media, and a new reference for it – the creative work ontology.

2 Methodology

In this paper, we adopt content analysis as the research method. We firstly review the literature to define the digital artifacts of new media. Then, we explore the characteristics of the digital artifacts and their relations with cultural products. Lastly, we draw our conclusions about the generalization of new media based on the creative work ontology.

3 The Digital Artifacts of New Media

3.1 Definition

In the paper, we refer to the definition of digital artifact by Jung and Stolterman [1] that a digital artifact is an "existing as well as potential [type] of physical product that delivers digital contents through its interactive features." This definition covers any mobile phone, desktop computer, laptop computer, tablet, wearable device, etc., with or without an Internet connection, that allows interactivity to trigger the delivery of digital content in the form of music, video, text, graphical information and so on. Any of these physical products can be a medium. We can collectively name all these physical products as new media empowered by our latest technologies. Brummett and Duncan [2] depict the relation between medium and technology as below:

> "Any medium is an extension of the human sensorium before it is a technology or a means for exchanging information. Many different technologies might be considered the same medium if they extend the sensorium in similar ways, as illustrated in Altheide's [4] insightful claim that keyboards – including computers, the telephone, food processors, calculators, and video games – are all the same medium insofar as they extend the sense of touch in the same way, although clearly they are not all the same technology (pp. 207–230)."

When mixing the viewpoints of Brummett and Duncan with the definition of digital artifact by Jung and Stolterman, digital artifacts are also an extension of the human sensorium through their haptic designs–their key interactive feature. Since they extend the human sensorium in the same way, they are or represent the same medium collectively known as new media. "In each case, then, an artefact is seen as extending a part of the body, a limb or the nervous system. And, as far as McLuhan is concerned, these are media" [5].

Furthermore, the digital content of new media is ever changing and growing. They are made by either individual creations or co-creations with at least two individuals possibly including their user or viewer. Brummett and Duncan state that, "A medium is a process, not a thing; it is something that people do, not a tool that people use" [2]. The digital artifacts also reflect the production process of new media.

Note that we are not challenging McLuhan's viewpoint that understanding a medium has nothing to do with its contents but its power [5]. In fact, we rarely discuss any specific content of a digital artifact in this paper. What we are focusing on in this section is why digital artifacts (inclusive of their contents) can represent new media. We take Tipp-Ex's interactive videos published on YouTube in 2011 as examples [6]. One of the surprises of these interactive videos was that the main character could stretch out his hand to get the Tipp-Ex correction tape on the top right corner of a YouTube page to wipe out a word in the video title above the video player. A viewer could type in a word into that "blank" area to change the ending of the original story. This unique feature of Tipp-Ex's interactive advertising campaign on YouTube required additional programming by YouTube and therefore a special cooperation between it and Tipp-Ex. In this case, YouTube running on a specific digital device collectively named a digital artifact showed its capability as a new media. According to Lister's discussion on McLuhan's slogan "The medium is the message" [5], such a digital artifact was telling us:

> "the way that it extend[ed] and speed[ed] up forms of 'human association and action' ... how it then changed the meanings (remediated) of already existing technologies and the kinds of human organization built around them ... the change of scale or pace or pattern that it introduce [d] into human affairs."

YouTube equipped with Tipp-Ex's interactive videos – the contents of a digital artifact – enabled us to grasp the power of a new media.

3.2 The Characteristics of the Digital Artifacts

According to Hung [7], cultural products have aesthetic and utilitarian purposes, rules, structures, features, and biases. In fact, Hung and Choy [3] already emphasize that the features are the cultural product attributes that can be detected by our five senses while the structures are hidden and require our analyses to find them out, and the biases are the cultural preferences over a period of time. All of them are bounded by rules. An example is clothing fashion with fashion theory as the rule, fabric combination and cutting as the structures, color, weight and texture as the features, and the age group, social class and occupation of a target market as the biases [3]. On the other hand, the digital artifacts also have features that allow our senses to find them out easily. Examples are buttons, touchpads, illumination, and sound. The structures of the digital artifacts are also hidden and require our analyses to locate them. Examples of structures are system design and hardware components. Moreover, there are different versions and even generations of these digital artifacts that cater for different users, forming the biases of these digital artifacts regarding user preferences and user experiences. Similar to cultural products, the features, structures, and biases of the digital artifacts are also bounded by rules – cultural rules, physics, ergonomics, etc. In addition, the study by Jung et al. [8] reveals that a digital artifact possesses four types of properties, namely, the physical aspect, the functional aspect, the informational aspect, and the interactive aspect. The physical one includes the aesthetic qualities of a digital artifact offering emotional values. The functional property embraces features. In short, digital artifacts and cultural products do share the same set of characteristics – rules, structures, features, and biases.

In fact, the availability of the rules, structures, features, and biases of the digital artifacts can also be found in the discursive interface analysis. Lamerichs and Te Molder [9] claims that the main characteristics of the discursive approach include (i) an emphasis on its relationship with certain social phenomenon; (ii) analyzes of text and talk as social practices and interactions leading to social actions (the text and talk can be embedded in new media), and (iii) a focus on participant involvement for ongoing development in communication (through new media). Stanfill [10] further proposes discursive interface analysis to study "websites' embedded assumptions about their own purpose and appropriate use." It emphasizes the "acceptance or rejection occurs within pre-determined options" through three types of affordances – sensory affordances, functional affordances, and cognitive affordances.

Sensory affordances focus on the design choices of websites that enable users to sense websites and therefore "[apportion] scarce attention to both reflect and reinforce [the] assumptions and valuations" of websites [10]. These affordances include color, text, images, sounds, motion, placement, and interactivity. Functional affordances construct the norms of websites defining what they can do in general. Lastly, cognitive affordances focus on the linguistic and nominative features of websites informing users of their capabilities through names, labels, and classifications – "social act of meaning-making." Stanfill highlights that his affordances of websites are about what the websites can furnish and offer to their users. Discursive interface analysis is also an understanding of "cultural common sense about what users do (and should do)."

From our perspective, sensory affordances concern the features of digital artifacts while functional affordances are about the rules and structures of the digital artifacts that support the appearance of the features embracing social constructs and net neutrality. Furthermore, cognitive affordances are the biases representing cultural constructs. The characteristics of the digital artifacts and the three affordances of the discursive interface analysis are interrelated to a certain extent.

Indeed, Lamerichs and Te Molder [9] already states that the discursive approach targets "a much more dynamic view of identities and social norms." The social norms are rules on the basis of which different identities are formed including copyright and fair use issues [11]. In a way, social norms are like cultural rules that are stable over a period of time, while identities are changeable like the features of cultural products.

In addition, there are studies that reveal the indivisibility between digital artifacts and cultural products. For example, Jung and Stolterman [1] stress that digital artifacts also possess aesthetic and functional qualities, and "now physical and digital qualities of computational materials are not separately considered anymore, but tend to be explored more holistically with focus on aesthetic qualities of a design." Moreover, Hughes and Lang [12] specially describe non-digital cultural products as fixed assets and digital cultural products (inclusive of the digital artifacts in new media, from our point of view) as liquid for transmutability margin.

4 The Creative Work Ontology

Hung and Choy [3] claim that cultural products can be described by their creative work ontology consisting of rules, structures, features, and biases as the ontological properties. The creative work ontology supports the design of a creative system equipped with computational creativity for cultural product creation.

We argue that since digital artifacts and cultural products share the same set of characteristics, and the creative work ontology can properly describe cultural products, it is possible to use the creative work ontology as the foundation to explore the generalization of new media.

5 The Generalization of New Media

In this paper, we contend that the generalization of new media carries the below proposition with reference to the creative work ontology:

The generalization of new media consists of rules, structures, features, and biases.

Below is a case study about the feasibility of applying the above proposition to the designing of Facebook's visibility.

5.1 A Case Study on Facebook

Facebook maintains all the necessary structures of a website that allow high traffic. When it was launched, its features quickly made it a unique social media platform for participant involvements. These features include the ways it organizes digital images, digital videos, text information, and notifications. From our perspective, one of the biases that shaped the features was the need for organizing multimedia information in a much more user-friendly manner. One of its rules is about the protection of privacy and therefore, only accepted friends can view a person's profile.

Bucher [13] stresses that the success of Facebook has directed software designers to consider visibility when designing software. Herein, visibility refers to the audiovisual elements that we can have on Facebook. The new modalities of visibility on Facebook caused by the EdgeRank algorithm have put "the medium specificity of the architectural organization of visibility, i.e. what can be seen and heard, to a large extent a question of software." Bucher further states that:

> "For instance, how do the 'Open Graph protocol', 'Like Button' and third-party apps figure into the articulation of different modes of becoming visible or invisible? How and to what extent different software components and code interact and cooperate in the production of social relations and subjectivity become critical questions for further research into Web 2.0 platforms."

Bucher's questions form an interesting insight into the design of Facebook that sheds light on the use of the proposition for better visibility. That is, it is possible to define the features on Facebook supported by the appropriate structures to offer a proper level of visibility. These features include the design of the Like Button and the adoptions of the third-party apps. The structures include software design, software components, codes, and functional affordances. Moreover, the biases relate to the influences of the social relations, subjectivity, and even subject position over the features. The new combinations or rearrangements of the rules, structures, features, and biases of Facebook give rise to better visibility.

6 Conclusion

In this paper, we work towards a generalization of new media and conclude our findings with a proposition. That is, the potential use of rules, structures, features, and biases for the generalization. This paper also provides a case study about applying these four potential properties to improving the visibility of Facebook. In the future, we will further solidify our proposition and develop a new media ontology.

References

1. Jung, H., Stolterman, E.: Material probe: exploring materiality of digital artifacts. In: Proceedings of the Fifth International Conference on Tangible, Embedded, and Embodied Interaction, pp. 153–156. ACM (2011)
2. Brummett, B., Duncan, M.C.: Toward a discursive ontology of media. Crit. Stud. Media Commun. **9**(3), 229–249 (1992)
3. Hung, E.C.K., Choy, C.S.T.: Conceptual recombination: a method for producing exploratory and transformational creativity in creative works. Knowl.-Based Syst. **53**, 1–12 (2013)
4. Altheide, D.L.: Media Power, vol. 158. Sage Publications Inc., Thousand Oaks (1985)
5. Lister, M.: New Media: A Critical Introduction. Taylor & Francis, Abingdon (2009)
6. TIPPEX 1 case study. https://www.youtube.com/watch?v=YhrWA9ET9us
7. Hung, E.C.K.: Tackling design fixation of cultural product designers through homeomorphism. In: Chung, W., Shin, C. (eds.) Advances in Affective and Pleasurable Design, pp. 491–498. Springer, Cham (2016)
8. Jung, H., Stolterman, E., Ryan, W., Thompson, T. Siegel, M.: Toward a framework for ecologies of artifacts: how are digital artifacts interconnected within a personal life? In: Proceedings of the 5th Nordic Conference on Human-Computer Interaction: Building Bridges, pp. 201–210. ACM (2008)
9. Lamerichs, J., Te Molder, H.F.: Computer-mediated communication: from a cognitive to a discursive model. New Media Soc. **5**(4), 451–473 (2003)
10. Stanfill, M.: The interface as discourse: the production of norms through web design. New Media Soc. **17**(7), 1059–1074 (2015)
11. Srinivasan, R.: Re-thinking the cultural codes of new media: the question concerning ontology. New Media Soc. **15**(2), 203–223 (2013)
12. Hughes, J., Lang, K.: Transmutability: digital decontextualization, manipulation, and recontextualization as a new source of value in the production and consumption of culture products. In: Proceedings of the 39th Annual Hawaii International Conference on System Sciences, vol. 8, p. 165a. IEEE (2006)
13. Bucher, T.: Want to be on the top? Algorithmic power and the threat of invisibility on Facebook. New Media Soc. **14**(7), 1164–1180 (2012)

Communication of Design: Applications

Calligraphy Practice App, System and Tool for Learning Calligraphy

João Brandão(✉)

CIAUD Centro de Investigação em Arquitetura, Urbanismo e Design,
Faculdade de Arquitetura da Universidade de Lisboa,
Rua Sá Nogueira, Pólo Universitário, Alto da Ajuda, 1349-063 Lisbon, Portugal
joao@b2design.pt

Abstract. This research project aims to create a functional system for learning and practicing flat tip calligraphy where students can acquire calligraphic skill independently without the aid of a teacher. The process of learning calligraphy involves learning the gestures and right movements while controlling nib inclination to obtain thick and thin areas in the correct sections of the character. There are two main skills to be acquired. One is the memorization of all the gestures necessary to draw the characters of a certain style. Two, the development of a conscious hand dexterity that will deliberately produce the necessary thick and thins by control of nib inclination in the movement. We are currently developing the Calligraphy Practice app, a system that has been proven to fulfill the previously mentioned needs. For this, we created a digital tool that emulates a flat tip brush accurately by allowing full control of the inclination throughout the entire drawing process.

Keywords: Calligraphy · Education · Software development · Mobile app

1 Introduction

Developing, or rather *re*-developing, an art involves *the tracing of in one's own experience of a process resembling its past development.* (Johnston 1994, p. xx)

Inventions and innovations spring from the need for a certain product that is not yet available or simply fails to fulfill our expectations. Unsatisfied with existing software for calligraphy, we set out to develop a more adequate system for learning and practicing calligraphy.

In this paper, we are presenting the Calligraphy Practice app, which includes the development of original tools to fulfill its purpose.

The inventions have been registered at the United States Patent and Trademark Office which granted the patent with Publication number US20140184546 A1.

The Calligraphy Practice app is currently available to the public and is complete system with several ways to learn and practice seven different styles of calligraphy.

The app was tested and validated as a viable self learning system, by method of focus groups from ages 8 and up.

A.G. Ho (ed.), *Advances in Communication of Design*, Advances in Intelligent Systems and Computing 609, DOI 10.1007/978-3-319-60477-0_11

2 Problem

The reason why today's type and fonts look like they do is directly related with their historical antecedents. Calligraphy has played an important part in the evolution of font design.

The shape of today's type and fonts is due to their own particular history.

Typography is derived from calligraphy, which has always been strongly related to the tools, materials and methods. Western calligraphy was mostly developed since ancient roman times until the invention of the printing press by Gutenberg. That was the era of scribes that copied by hand the books at monasteries.

The tools used were pens or quills with a flat tip. Because of this feature as they wrote, the characters would naturally assume thin and thick areas.

With the invention and massive use of typography, this practice is lost and font design steers away from calligraphy. That is until Edward Johnston (1872–1944) researches and rediscovers calligraphy and publishes the book "Writing & Illuminating & Lettering" in 1906.

"Johnston's most important insight was that the forms of our letters grow naturally from the use of the edged pen—in other words, the tool determines, to a large extent, the way the writing will look." (Calderhead 2012, p. 3)

This event will change forever the font design principals of the 20th and 21st century. New fonts appear with refined contrasts between thick and thin that strongly enhance their legibility and personality in both serif and sans serif fonts.

Nowadays every serious font designer has knowledge of calligraphy. This gives them an understanding of the basis of font anatomy, the reasons for its contrasts and structure.

One can read about calligraphy, but it's with practice that one develops a deeper understanding of a character's anatomy necessary for the design of new typefaces.

Without living the experience of properly handling the calligraphic flat brush or flat tip pen and learning to feel the characters unravel in different ways, there is no actual learning benefit.

"To reach that level of ease, you need to set aside some time in your schedule to simply write. Hour by hour, day by day, you train your hand so the motions become second nature. You build up muscle memory; you internalize the rhythm of writing. Calligraphy is like any physical skill: The dancer learns her craft by dancing, the pianist by playing scales, the soccer player by working out on the field with the ball. The calligrapher learns by spending time at the writing board." (Calderhead 2012, p. 1)

Today's design and illustration software often contain tools that function as calligraphic brushes.

These work by defining a fixed width and angle. This way it's possible to draw calligraphy and edit the shape.

The problem with these tools is that they don't allow the liberty of a real flat brush or flat tip pen.

With a flat paintbrush one can start to draw a character with one inclination and finish with a different one; it's possible to press harder or lighter to achieve a different thickness along the same stroke; one can lift and turn the paintbrush to finish off serifs;

all this can be done in a natural instinctive way that works as an extension of the hand, focused only on the result.

There are two different kinds of nibs: italic and flex nibs. The writing tip of an italic nib is cut straight across and thinned top to bottom. Line variation is limited to thick or thin. Vertical up and down strokes are thick and horizontal sideways strokes are thin.

As to flex nibs, line variation isn't caused by tip shape as much as it is due to the pressure applied while writing. The tip is round and the tines are longer than other non-flex nibs. Tines curve up-ward and spread apart as pressure is applied during the writing process. The wider the space between the tines, the thicker the lines become.

This liberty is key, not only it allows mistakes that are very important on the educational process, it is with this liberty that the experts finish off their finest details.

The existing software is in my opinion is inadequate for the educational process and inefficient for the expert.

Expert calligraphers will always do better calligraphy and substantially faster when done by hand. The trouble of adjusting thickness and angle between each stroke is extremely time consuming, and the inability to rotate the brush as one is drawing is a big fault.

3 Objective and Solution

The objective of this project was to develop a tool that would solve all of those problems and go beyond by benefiting of features of digital media like post editing and undo, in a simple fast and efficient way.

The main part was to create a tool that would return this liberty to the user, in a way that would be easy and instinctive, removing the interface buttons and switches by returning all the flat brush control to the user's hand and the hand only.

4 How It Works

An opportunity was found in the multi-gesture software of tablet computers and software for allowing the use of multiple fingers.

The simple placement and sub sequential drag of those fingers will feed coordinates for the system to plot/draw a simulation of what will look a lot like a calligraphic flat brush.

The distance between fingers will determine the thickness of the stroke, and their relative position will determine the inclination or angle.

As the fingers drag over the screen a shape is formed… Or digitally finger painted.

This way, because it works without the need to set parameters previously and individually like in traditional software, and because the user can change the relative position of his fingers as he draws, the user is painting with all the liberty and instructiveness of a real brush.

This system allows for both change in rotation and change in thickness simultaneously and all of this can be done during the drawing gesture just like it was done during the brush stroke.

This system was developed using programming language adequate for the iPad. Although any known programming language may be used to practice the present invention, the system was created using x-code in objective C. The following is an explanation of selected code, present in the patent registration:

To generate a brush with a line, we used three sets of two touch points as shown in the function midPoint returns the middle point between two points.

drawCurvedFigureWithPrevious draws a geometric figure of two bezier curves and a straight lines that corresponds to a segment.

```
BEZIER_PERCENTAGE is a constant for tweaking the smoothness of the curves.
- (void)drawCurvedFigureWithPrevious:(TouchPoints *)previous current:(TouchPoints
*)current next:(TouchPoints *)next inContext:(CGContextRef)context {
  Point mid1 = midPoint(previous.start, current.start);
  Point mid2 = midPoint(current.start, next.start);
  Point mid3 = midPoint(current.end, next.end);
  Point mid4 = midPoint(previous.end, current.end);
  ContextBeginPath(context);
  ContextMoveToPoint(context, mid1.x, mid1.y);
  Point cpx1 = pointWithPercentage(BEZIER_PERCENTAGE, mid1, current.start);
  Point cpx2 = pointWithPercentage(BEZIER_PERCENTAGE, mid2, current.start );
  ContextAddCurveToPoint(context, cpx1.x, cpx1.y, cpx2.x, cpx2.y, mid2.x, mid2.y);
  ContextAddLineToPoint(context, mid3.x, mid3.y);
  Point cpx3 = pointWithPercentage(BEZIER_PERCENTAGE, mid3, current.end);
  Point cpx4 = pointWithPercentage(BEZIER_PERCENTAGE, mid4, current.end);
  ContextAddCurveToPoint(context, cpx3.x, cpx3.y, cpx4.x, cpx4.y, mid4.x, mid4.y);
}
```

To draw each of the segments between the points, the routine is called every time the view is drawn. It iterates over all the touch points recorded and calls.

```
drawCurvedFigureWithPrevious:current:next:inContext:.
- (void)drawView
{
 ContextRef context = UIGraphicsGetCurrentContext( );
 for ( Array *p in _paths)
 {
   //Enumerator for all the sets of 2 touches.
  Enumerator *touchesEnumerator = [p objectEnumerator];
   //TouchPoints is a class that stores two points rep-
resenting the two
   finger touches.
  TouchPoints *next;
  TouchPoints *current = [linesEnumerator nextObject];
  TouchPoints *previous = current;
  while (next = [linesEnumerator nextObject]) {
   [self   drawCurvedFigureWithPrevious:previous   cur-
rent:current
 next:next inContext:context];
   previous = current;
   current = next;
   ContextClosePath(context);
   ContextDrawPath(context, Fill);
  }
 }
}
```

To smooth the swath, the drawn points are not necessarily equal to the touch points and a scaling is applied given by a percentage number.

```
  Point pointWithPercentage(Float percentage,Point
p1, Point p2)
  {
   return PointMake(p1.x + (p1.x - p2.x) * per-
centage, p1.y +
   (p1.y - p2.y) * percentage);
  }
```

5 The Calligraphy Practice App

Being this developed on a digital platform, the final product was completed with several features and functions so it would function as a complete learning package.

One important feature is that the application can be used for practice everywhere without having the inconvenience of having to carry around inks, paper, books and other materials essential for the practice of calligraphy.

Instructions to several calligraphy styles have been designed as instructional trace over base, specifically for this application:

- A foundation hand, based on Caroline calligraphy that is the base of roman type;
- capitals;
- an italic hand;
- A textura quadrata style;
- A fraktur style;
- Gothic capitals;
- Uncials;
- Numbers and punctuation signs.

Other functionalities are:

- The ability to undo;
- The ability to edit just one character in a word;
- The ability to email the document as vector drawing for further editing and fine tuning with professional software;
- Allows to trace over a base instructional font especially developed for the purpose;
- Turn on and of guides for the baseline, x-height, ascenders and descenders;
- The ability to learn alone;
- Connect to a projector and teach, demonstrate how to draw a specific character;
- Scale and print in any size;
- Easily share the work;

This project is currently fully functional and available to the public in over 150 countries worldwide thru the app store.

With its implementation, tests have been made with several types of end users, from school children ages 8 and up, to university students of graphic design and typography which validated the software's efficiency.

Acknowledgments. 1. CIAUD Centro de Investigação em Arquitetura, Urbanismo e Design,
2. Faculdade de Arquitetura da Universidade de Lisboa.
3. FCT Fundação para a ciência e Tecnologia, Portugal.
4. Vasco Conde, Elastic Fiction.
5. Stephen J. Driscoll of Saul Ewing LLP.
6. Kelly Morgan of Saul Ewing LLP.

References

Calderhead, C.: Calligraphy Studio: The Ultimate Introduction to the Art of Hand Lettering. Sterling, New York (2012)
Harris, D.: The Art of Calligraphy. Dorling Kindersley, London (1995)
Johnston, E.: A Book of Sample Scripts. Victoria and Albert Museum, London (1966)
Johnston, E.: Writing & Illuminating, & Lettering. A & C Black, London (1994)
Noordzij, G.: The Stroke: Theory of Writing. Hyphen Press, London (2005)
Waters, S.: Foundations of Calligraphy. John Neal Bookseller, Greensboro (2006)

Smart Device-Based Notifications: A Survey on User's Satisfaction of Traditional Notification Mechanisms

Shirley Yoong, Gustavo López(✉), and Luis A. Guerrero

Universidad de Costa Rica, San José, Costa Rica
{shirley.yoong, gutavo.lopez_h, luis.guerreroblanco}@ucr.ac.cr

Abstract. Notifications represent a key element in smart devices. In fact, a device could be considered "smart" if it is able to provide different type of messages or notifications to the users. In this way, people who use smart devices receive a significant number of notifications daily. However, how many notifications can a user receive and attend daily? If the number of notifications increases a lot, it is possible that the user experience decrease. In this article, a study is carried out on the level of user's satisfaction related to the number of notifications received in different smart devices. To achieve this goal, a survey was applied taking into consideration the following devices: smart speaker, smart fridge, smart TV, smart watch, smart phone, and car's integrated computer system. The study comprises the participation of 439 people from 8 countries.

Keywords: Smart devices · Notifications · User experience

1 Introduction

Smart devices are electronic devices that have computational capabilities and are connected to a network. Smart devices are being rapidly adopted; everyday new devices and applications are released [1]. Some examples of smart devices include: smart phones, smart watches, smart bands and key chains, smart fridges, and smart speakers (usually used to access intelligent personal assistants), among others.

Several concepts are related with smart devices including but not limited to: Ubiquitous Computing (Ubicomp) and Internet of Things (IoT).

In Ubicomp field, computational devices appear to be available everywhere and at any time, in an invisible and seamlessly way [2]. The implementation of Ubicomp systems requires a change in the traditional system design and engineering paradigm. Moreover, the development of new technologies, microprocessors, sensors, networking protocols and materials is required to achieve a real application [3].

In IoT field, the devices promoted by Ubicomp must be connected to a network. IoT systems allows those devices to be sensed and controlled remotely, either by users or by third party applications [4]. Several efforts are conducted to develop sustainable and secure IoT solutions. However, most of these efforts are purely technical (e.g., protocol standardization, and development guidelines, among others).

A.G. Ho (ed.), *Advances in Communication of Design*, Advances in Intelligent Systems and Computing 609, DOI 10.1007/978-3-319-60477-0_12

IoT and Ubicomp were strongly enforced by the development of small and inexpensive technologies. Many of the applications developed in these areas include sensors. In these cases, the traditional use of smart devices has been to gather information and allow applications to use it (e.g., know where the user is, know what the user is doing, and tracking animals). It is noticeable that one of the functionalities of these devices is that they receive different kind of notifications [5].

Currently, the development of more sophisticated and complex devices does not only allow to gather user data but also to provide them with "valuable" information. We denote valuable because this concept is very ambiguous and depends on every user (i.e., a message valuable to one user may not be for another).

In addition, people have introduced smart devices in their daily activities and, in consequence, receive a certain number of messages daily. However, if the number of messages increases the user experience could decrease. It is unclear the relationship between people's satisfaction with the amount and priority of notifications they receive on their smart devices.

This paper presents the results of a survey in which participants were asked about some smart devices, their willingness to use them and receive different types of notifications in some of those smart devices. The purpose of this research was to characterize the way in which people would prefer to get notifications via smart devices.

The survey was conducted during May 2016. A total of 439 respondents from eight countries participated in the survey. The country with more responses was Ecuador (46.8%) followed by Costa Rica (21.5%), Venezuela (10%), Colombia (9.7%), Peru (5.7%), USA (3.1%), Mexico (1.6%), and Spain (1.6%).

The survey was designed to assess the user's familiarity with technological devices, the types of notifications they will be willing to receive in a set of smart devices, and their preferences on those devices.

The assessed smart devices include: (1) smartphone, (2) smart watch, (3) smart fridge, (4) smart TV, (5) a car's integrated computer system, and (6) smart speaker. These categories were deliberately selected to represent different types of smart devices (e.g., handheld, wearable, and embedded).

The results presented in this paper are part of a larger research project. The project goal is to avoid the excessive and inordinate use of notifications that might be overwhelming for users. Furthermore, the project seeks to allow the proper design of notifications to reduce user's cognitive efforts.

The rest of this paper is structured as follows: Sect. 2 describes the generation of notifications. Section 3 presents the methodology used for this study and its participants. Section 4 analyzes the survey to verify the user's satisfaction. Finally, Sect. 5 presents the conclusions of the study and the further work.

2 Related Work

Nowadays, smart devices use different ways of reporting so people are awake of many activities that they work up daily [6].

Regarding the computational domain, several algorithms are used to solve problems related to the notifications received by people. These algorithms consider the users' needs. The notifications are managed by the most suitable intelligent devices according to the notification that is received, in order to achieve less intrusion in the daily life of the people [1].

In a previous works, a study on people's preferences for receiving notifications was conducted. Authors state that the notifications received on smart devices are variables and they depend on the functionality of the application. In addition, it is affirmed that smartphones are the most favourite devices for the people [7, 8].

Another work that has been developed to cover the users' needs in the way they receive notifications on their smart devices is the meS-chup IoT platform. In this platform, when a notification arrives to the smart phone, the system redirects it to different intelligent devices in the environment [6].

Another study makes a comparison between smartphones, smart watches and smart TVs. This study indicates that smart TVs are devices that can be viewed at the same time by several people so the mechanism of notifications should consider privacy and should respect the attention of other people. This situation does not happen with the other two devices mentioned above because they are personal devices [9, 10].

It should be considered that the notifications are not only used on a personal level otherwise in a professional level. In the professional domain must be considered the priority of the tasks performed by the people and, in case of carrying out shared tasks, to notify in the correct time of the different stages of the work. Both notifications (personal and professional) have to be directed in the best way, so the users do not lose concentration in their work [11, 13, 14].

In order to implement solutions that cover people's needs in relation to the different notifications they receive, the environment that surround them must be considered, not only the devices but also the accesses that people need to receive their notifications on time.

In case of online cooperative work and their corresponding notifications, people must have a participatory attitude that accomplishes the assigned tasks [12].

3 Methodology

This section presents the procedure followed to conduct the research reported in this paper. First, we will briefly describe the considered smart devices. Afterwards, we will describe the survey participants. Finally, we will present an overview of the questionnaire used to gather the participant's expectations and realities on smart devices as notification mechanisms.

3.1 Assessed Smart Devices

Speaker. Smart speakers are hands-free devices and voice activated interface, which allows the user to interact with Internet based services (usually personal assistants) [15, 16]. Some examples of smart speakers are the Amazon Echo and Google Home.

Smart Fridge. A Smart Fridge is a technology-enabled refrigerator programmed to sense the products being stored inside it. The newest additions to Smart Fridge functionalities are touchscreens in the doors that can show any type of information. An example of a Smart Fridge is Samsung's Family Hub [17].

Smart TV. Smart TVs are television sets with integrated Internet features and connectivity. The main functionality of Smart TVs is to supply contents (e.g., photos, movies, music) that traditional televisions were not able to deliver [18].

Smart Watch. Smart watches are wearable pieces of technology with Internet access (traditionally using a smartphone as middleware) that use a graphical display and a touchscreen to deliver content to users [19].

Car's Integrated Computer System. Nowadays most vehicles have integrated computers. However, some of the most luxurious cars have user interfaces that have most of the functionalities of a traditional computer. By car's integrated computer, we refer to that user interface that allows direct interaction with the user.

Smartphone. Smartphones are handheld devices with multi-purpose operative systems. Again, these devices offer almost all the functionalities available in traditional computers. They are also equipped with several sensors that allow data gathering.

3.2 Participants

Participants were recruited through social networks and e-mail. Any person who wished to participate was allowed to do so. Of all 439 participants, 52% (229) were men and 48% (210) were women. The countries from which the survey was answered were the following: USA, Mexico, Costa Rica, Colombia, Venezuela, Ecuador, Peru, and Spain. Figure 1 provides more details of the participant's geographical and gender distribution.

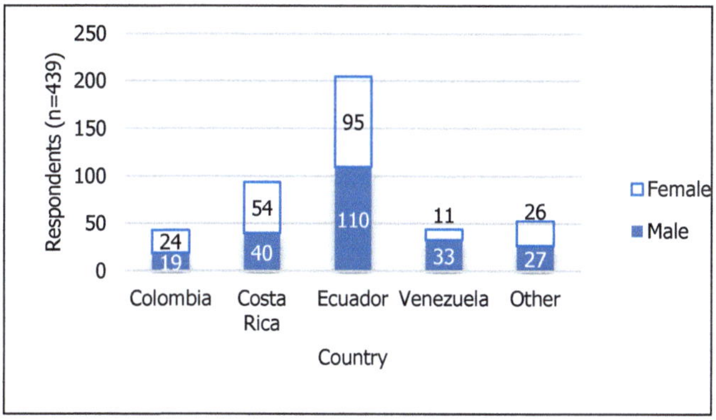

Fig. 1. Participant's distribution by country and gender

Participant's age behaviour shows a bimodal distribution with humps on people ranging from 21 to 25 and people from 41 to 45. Figure 2 shows this data. Even though we do not intend to generalize the results gathered in this research, the sample is appropriate and age distribution at least would allow us to believe that results gathered of people ranking 16 to 50 are representative of the population.

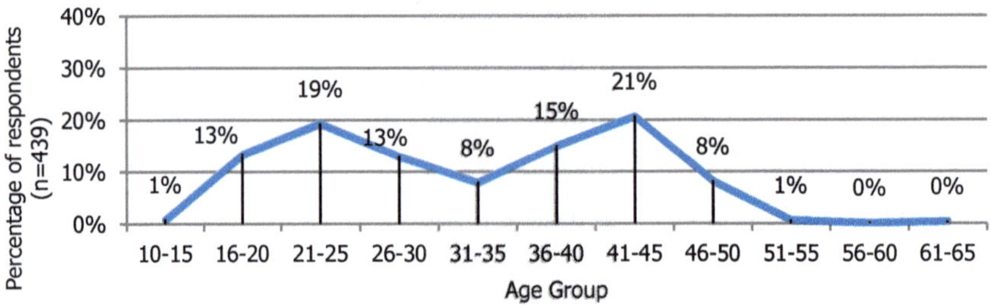

Fig. 2. Participant's distribution by age

The participant's familiarity with smart devices was assessed. Table 1 shows the percentage of respondents that were familiar with each device. Even though the Smart speakers are quite new in the market, 39% of the participants knew or heard about it before this study.

Table 1. Participant familiarity with smart devices

Smart device	Knows	Do not know
Speaker	39%	61%
Frige	55%	45%
Car	79%	21%
TV	86%	14%
Watch	90%	10%
Phone	100%	0%

The participants were asked about how many hours per day they spent in contact with technological devices. The answers available were arranged in a Fibonacci sequence. Figure 3 shows the distribution of the amount of hours spent by the

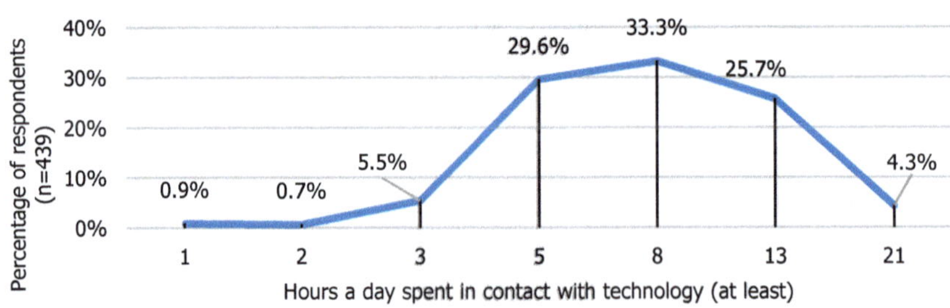

Fig. 3. Hours a day spent by participants using technological devices

respondents using technological devices. The expected answer was at least 8, since nowadays most people spent at least their working hours using technology. However, the negative skew of the distribution is significant; 25% of the respondents spent at least 13 h a day using technology and more than 4% use it more than 21 h a day.

3.3 Questionnaire

The survey conducted in this research had 17 questions, divided in 3 sections. The first section was defined to gather the sample demographic information. Questions regarding country of residence, sex, age, academic degree, civil statues and employment status were also asked.

The second section was designed to assess the participant's attitudes towards technology in general and towards smart devices. A set of questions regarding technology perspectives was used. The amount of hours the participant spends using technological devices and the amount of notifications per day that they receive were also addressed. A Likert scale question about user satisfaction was included in this section.

Finally, the third section addressed smart devices. Participants were asked which type of notification they would allow in each of the assessed devices. Also, they were asked to rank the six devices in the order in which they will probably review them. To conclude the survey, four senses of urgency (e.g., informative message, warning, alarm and critical alarm) were introduced. Users were asked about each type of alarm and which devices they would prefer for deliver this kind of notification. The next section describes the main results gathered from our survey and the analysis performed.

4 Results

As Table 1 showed that most participants were familiar with smart devices. However, in case some of them were not familiar, the survey provided videos that exemplified the use and characteristics of each of the assessed smart devices.

Figure 4 shows the distributions of answers to the question: which device are you more likely to see? This question clearly states that respondents should assume that

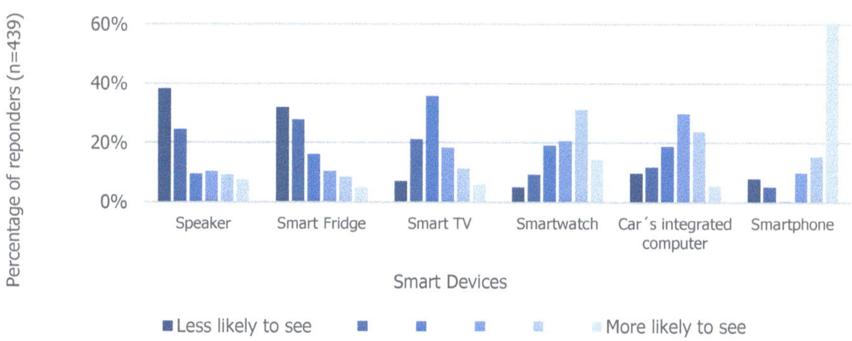

Fig. 4. Participant's response about which device they are more likely to see

they have all the available devices. Both the Amazon Echo and the Smart Fridge show a positive skew towards the less likely to see. Therefore, they were the worst ranked devices.

The TV, the smart watch and the car's computer screen seem to have a normal distribution. Finally, the smartphone shows a negative skew distribution towards the more likely to see. Therefore, the smartphone was considered by the respondents the most likely to see device in case a notification is delivered.

In a quantitative way, to the least frequent to the most frequent revision in relation to the notifications delivered, the result obtained in the survey was: 38% for the smart speaker, followed by the smart refrigerator with 28%, smart television with 36%, then the vehicle with screen and integrated computer with 30%, smart watch with 31%, and finally the smartphone with 61%.

We asked participants how many emails they received each day. Again the Fibonacci sequence was used to provide participants with a tool for a good estimation. Figure 5 show the distribution of emails received each day and the percentage of respondents for each category. Even though, in Fig. 5 the growing of the distribution is not very steep, the majority of participants receive at least 13 emails per day.

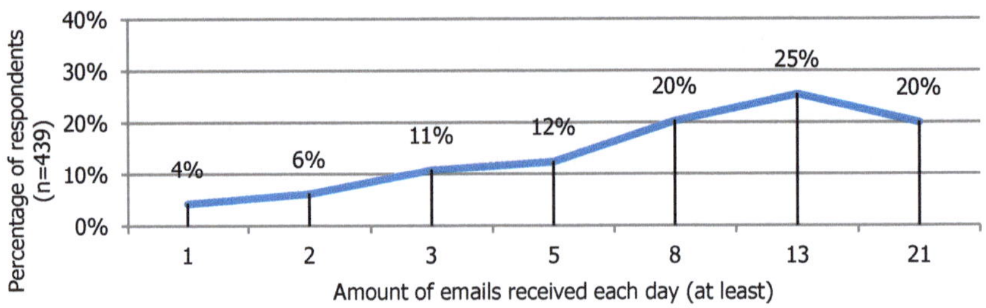

Fig. 5. Amount of emails received by participants

We asked participants to know how satisfied they were with the amount of notifications they got through their phones. Figure 6 shows the responses to this question. The level of satisfaction is positive, (i.e., 44.3% has a positive attitude towards the

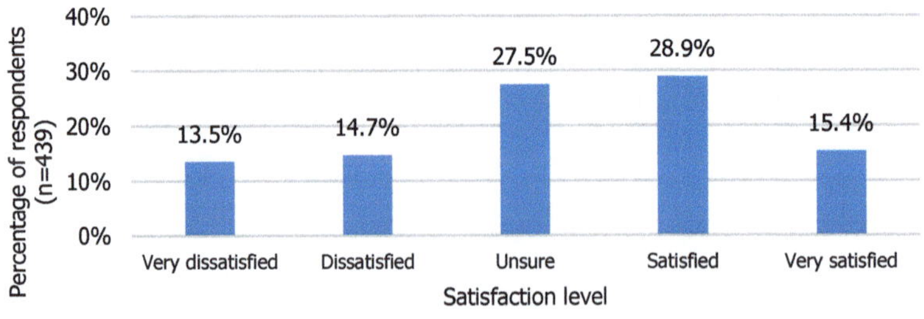

Fig. 6. Participant's satisfaction with the amount of notification delivered through their smartphone

amount of notifications received through their smartphone). However, almost 30% of the respondents declared that they are somehow dissatisfied with the amount of notifications. The rest (27.5%) is unsure.

Our first discussions on this topic lead to a combination of factors to see if correlations appeared. Figure 7 shows that the most dissatisfied respondents receive more than 34 emails per day. Moreover, users' satisfaction about the number of emails received in a day shows that, if the number of emails is minimal the satisfaction of the participants is very high 57% between satisfied and very satisfied. Otherwise, when the number of emails is greater or equal to 34 the satisfaction level only reaches 36%, 37% are unsure and 27% are very dissatisfied.

In the same graphic it is observable that the lowest dissatisfaction level (5% very dissatisfied) is when people receive two emails per day against 64% between satisfied and very satisfied. According to this result, receiving two emails per day would be the best option for people respondents.

Figure 7 also shows that when people receive 21 e-mails per day, the dissatisfaction level is the highest (37% between unsatisfied and very unsatisfied) and the satisfaction level is the lowest with 33% between satisfied and very satisfied.

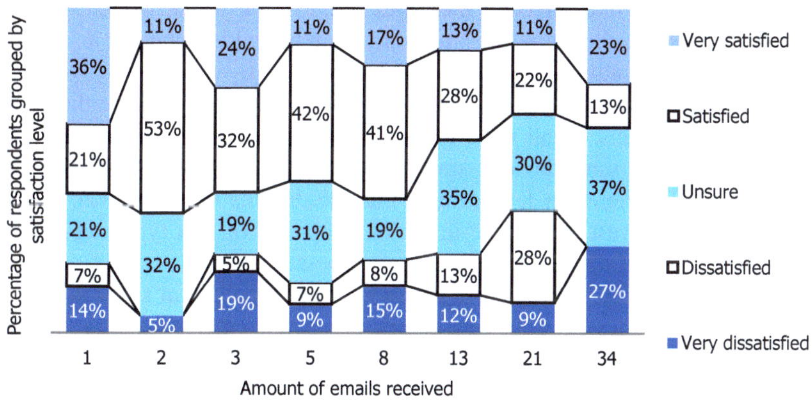

Fig. 7. Combination of satisfaction and amount of emails received per day

With respect to receiving three, five, eight and thirteen emails the satisfaction level (satisfied and very satisfied) is between 41% and 58%; and the dissatisfaction level (dissatisfied and very dissatisfied) is between 16% and 25%. In other words, respondents indicated that if they receive less mail it is better.

Figure 8 show that respondents like to receive messages with content public in smart watch, car's integrated computer system, and smartphone. In case to the home smart devices don't like to receive any notification.

We can indicate that the preference of the respondents in relation to receiving private messages is towards the smartphone with 69%, followed by the smart clock and vehicles with screen and integrated computer both with a 58%, the rest of the devices are with percentages less than 35%.

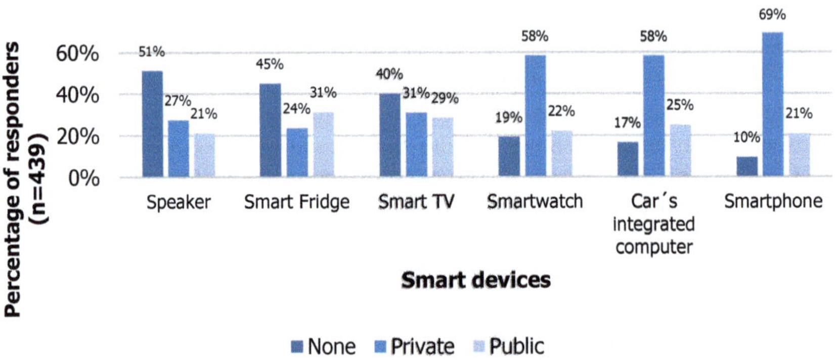

Fig. 8. Combination of the smart devices and the privacy of the messages

In terms of receiving public messages, respondents prefer smart cooling with 31% and other devices are in the range of 29% to 21% so we can consider that in all devices could receive this type of messages.

In home smart devices like smart speaker, smart refrigerator and smart TV, respondents would not like to receive messages with 51%, 45%, and 40%, respectively. In the case of other intelligent devices, respondents reported less than 20%, which is a minimal possibility of not receiving messages of any kind.

In Fig. 9, we can see the inclination of people by the type of message they want to receive on smart gadgets. In all three cases (less urgent, urgent and emergency message) people voted the same devices, smart phone and smart watch, which occupy the first two places of preference. It is evident that the devices smart speaker and smart refrigerator are the lowest percentage in the three options of sent type messages (less urgent message, urgent message and emergency messages).

In a quantitative way, according to the preference of people in smartphones, they would like to receive messages not very urgent, urgent and emergency with 22%, 36%, and 35%, respectively. For the vehicle with display and integrated computer the

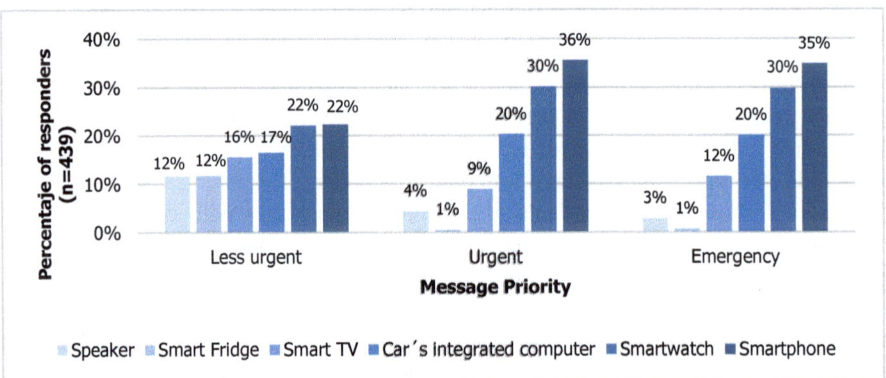

Fig. 9. Combination of smart devices and type of message

percentages are: 20%, 20%, and 17% that would like to receive emergency, urgent and not urgent messages, respectively. In the case of the smart watch, people's preference is 30% for emergency and urgent messages; and 22% for non-urgent messages. For home devices such as: smart TV, smart refrigerator and smart speaker to receive low urgency messages the percentage is 16%, 12%, and 12%, respectively. For urgent message their percentages are 9%, 1%, and 4%, respectively; and finally, for emergency messages have a percentage of 12%, 1%, and 3%, respectively.

Another interesting finding is that even though an extensive review of the responses was conducted, no differences could be determined when the responses were separated by age. In other words, the age is not a factor in the way people would like to receive their notifications.

5 Conclusion and Future Work

According to the answers obtained in the survey, it can be concluded that in the household devices (smart speaker, smart TV and smart fridge) people would not like to receive messages of any kind but in case notifications are received, they prefer that are not urgent. In the case of the smartphone, smartwatch and car´s integrated computer, they prefer to receive messages with private content and notifications can be of any of the three types raised in the survey (not urgent, urgent and emergency).

As for the amount of emails users receive, the bottom line is that the less you get the more satisfied the people will be. However, a common number of emails received lays between 8 and 21.

The most important result of this paper is that there are not significant differences between ages. Several analyses were performed categorizing the responses in different groups. However, no significant differences were found. Moreover, the gender categorization of the responses did not show significant differences either.

After analyzing the data, we can implement a decision-maker, according to the experience of the respondents, in order to better predict the delivery of the notifications assuming that they have different intelligent devices and that in turn embedded in your environment. The next step would be to implement a personalized system for delivery of notifications and thus achieve a less intrusion in the daily life of people.

Acknowledgments. This work was partially supported by the Centro de Investigaciones en Tecnologías de la Información y Comunicación (CITIC) Grant No. 834-B6-178, and the Posgrado en Computación e Informática (PCI), both at Universidad de Costa Rica.

References

1. Corno, F., De Russis, L., Montanaro, T.: A context and user aware smart notification system. In: User Modeling and User-Adapted Interaction, pp. 645–651 (2015)
2. Weiser, M.: The computer for the 21st century. Mob. Comput. Commun. Rev. **3**, 66–75 (1991)

3. Väänänen-Vainio-Mattila, K., Olsson, T., Häkkilä, J.: Towards deeper understanding of user experience with ubiquitous computing systems: systematic literature review and design framework. In: Abascal, J., Barbosa, S., Fetter, M., Gross, T., Palanque, P., Winckler, M. (eds.) Human-Computer Interaction, pp. 384–401. Springer, Cham (2015)
4. Visser, J. (ed.): Impact of M2M Communications and Non-M2 M Mobile Data Applications on Mobile Networks, Geneva (2012)
5. De la Cruz C.: Series Notification System: definición de un sistema de notificaciones personalizable porel usuario. Universidad Carlos III de Madrid, Departamento de Informática. http://e-archivo.uc3m.es/handle/10016/17184
6. Kubitza T., Voit A., Weber D., Schmidt A.: An IoT infrastructure for ubiquitous notifications in intelligent living environments. In: Ubicomp/ISWC (2016)
7. Moran, S., Fischer, J.E.: Designing notifications for ubiquitous monitoring systems. In: International Conference on Pervasive Computing and Communications, pp. 115–120 (2013)
8. Shirazi, A.S., Henze, N.: Assessment of notifications on smartwatches. In: International Conference on Human-Computer Interaction with Mobile Devices and Services Adjunct, pp. 1111–1116 (2015)
9. Vishnevsky, V., Andrey, L., Frolov, S.: Distributed computer and communication networks. In: International Conference on Distributed Computer and Communication Networks, pp. 11–27 (2014)
10. Weber, D., Mayer, S., Voit, A., Fierro, R.V., Henze, N.: Design guidelines for notifications on smart TVs. In: International Conference on Interactive Experiences for Television and Online Video, pp. 13–24. ACM (2016)
11. Paul, C.L., Komlodi, A., Lutters, W.: Interruptive notifications in support of task management. Int. J. Hum. Comput. Stud. **79**, 20–34 (2015)
12. Mulder, I.: Distributed, ambient, and pervasive interactions. In: International Conference on Distributed, Ambient, and Pervasive Interactions, pp. 566–574 (2014)
13. Telmo Neves, D., Santos, M., Pinto, M.: ReActOR: a middleware as a service to interact with objects remotely. In: International Conference on Industrial Technology, pp. 2433–2439 (2015)
14. Weber, D., Shirazi, A.S., Henze, N.: Towards smart notifications using research in the large. In: International Conference on Human-Computer Interaction with Mobile Devices and Services Adjunct, pp. 1117–1122. ACM, New York (2015)
15. Dempsey, P.: The teardown Amazon echo digital personal assistant. Eng. Technol. **10**, 88–89 (2015)
16. Amazon Inc.: Alexa Skills Kit. https://developer.amazon.com/public/solutions/alexa/alexa-skills-kit
17. Samsung: Family Hub Refrigerator. http://www.samsung.com/us/explore/family-hub-refrigerator/
18. Business Insider Inc.: What Is a Smart TV? http://www.businessinsider.com/what-is-a-smart-tv-2010-12
19. Pebble: Pebble. https://www.pebble.com

The Challenges of Online Meal Ordering Platforms: Website Design and Credibility, Customers' Trust, and Food Safety

Yangmin Zhang, Jieyu Wang(✉), and Jingyu Wang

International College Beijing, University of Colorado, Denver, Beijing 100083, China
{yangmin.zhang, jieyu.wang}@ucdenver.edu, jywangy2017@hotmail.com

Abstract. In modern societies using online meal ordering platforms becomes an emerging culture for Chinese citizens. The majority of the ordering platforms are based on smart phone applications. However, little research has been done on the credibility of online meal ordering applications and how to improve the design of these applications in China. Moreover, users of these applications begin to question about the credibility of these applications and trust issues appear. We investigate the challenges that these online meal ordering platforms face, for example, customers' trust and its application credibility. We send out online surveys to explore the reasons why these customers use these applications. We conduct interviews to investigate the reviews of these application users and the users' design suggestions for these applications. Observation has been conducted in order to better design the applications for these companies. The data show that the reasons why customers choose a meal ordering application are that 43.23% of the customers pay attention to various kinds of restaurants offered by the applications and 34.84% of them prefer low deliver fees. Well-designed application interfaces also attract 12.9% of the participants to use these applications. It also show that the participants concern about the information organization of the application design. Better navigation and information organization are expected for the platforms. We investigate the challenges that the online meal ordering platforms face, for example, website design, customers' trust, and application credibility. This paper aims to provide design guidelines for these ordering platform companies to improve their application credibility and gain trust from their customers.

Keywords: Online platform · Website design and credibility · Customers' trust · Food safety

1 Introduction

On-line meal ordering platform is a kind of new model cooperation between platform companies and restaurants, that platform companies provide online websites meal ordering service and deliver service, and restaurants provide food.

In the present society using on-line meal ordering platforms has become an emerging daily behavior for Chinese citizens. The majority of the ordering platforms are

A.G. Ho (ed.), *Advances in Communication of Design*, Advances in Intelligent Systems and Computing 609, DOI 10.1007/978-3-319-60477-0_13

based on smart phone applications, for instance Eleme, Meituan and Baidunuomi. On the one hand these platforms bring convenience to people who order meals, but on the other hand these applications hide the process of how chefs cook and what they cook from customers, and their designs sometimes are not suiting the customers' taste. This leads to using problems and bring food security problems which recently citizens care more about at present. Lots of small-scale manufacturers without business licenses sell their food though Eleme, and many of them have unsanitary cooking environment. This situation has been questioned by customers about the credibility of these applications and caused trust issues. We investigate the challenges that these on-line meal ordering platforms are facing, for example, website design, customers' trust and its application credibility.

There are three methods in the investigation: survey, interview, and observation. We decided to send out online surveys to explore the reasons why these customers use these applications. We conducted interviews to investigate the complaints from customers, the cooperation of the merchants and the food security department, and merchants' concerns from the platform company's perspective. The observation has been carried out in order to get advices for these companies to better design the applications. In the end, we gained useful data and specific feedback of the customers, staff of restaurants, and application companies. This paper aims to provide design guidelines for these ordering platform companies to improve their application design and credibility, and gain trust from their customers.

2 Literature Review

With the development of modern technology, applications of such meal ordering platforms bring a new culture to Chinese citizens. According to our research, academic documents and articles showed that three main parts are related to this topic: technology, brand and service, and food security.

In terms of technology, Fu et al. [1] analyzed users' comments and provided valuable views to help the app company's guidelines to revamp their apps. Yin et al. [2] showed a good app recommend model. Liu's [3] study proposed that we still can use the information interaction technology on the internet to supervise the food quality by ourselves. And it indicates that people's supervision has much impact on supervising the government to solve this kind of problem. Moreover, Gupta [4] suggested that we need develop this kind of technology, which may be helpful to the supervision of take-out package foods. It possibly may distribute different kinds of food to support people better services on foods.

In terms of brand and service, Gerson [5] emphasized that customer satisfaction and the quality of products are extremely important to the company. Swift [6] conducted more corporation try to put more information out into the public, which was a way to communicate with the people whose profit was relative with the corporation. Chen et al. [7] indicate that brand image significantly affected purchase intention.

In terms of food security, Boyce et al. [8] conducted two methods to make sure the security of food, origin labeling and traceability. Boyce et al. [8] did some research and express to the readers that packages is the most important element of takeout, and the soul of the sales takeout. The studies of Egana et al. [9] found that food hygiene training is limited by a lack of academic knowledge.

Taken together, the literature suggests that it is significant to focus on the qualification of a company's brand image, service; the usage of new technology, like applications and interaction technic; and food security. However, the specific content of an on-line ordering platform is missing. These studies do not mention that the security between the food producing process, and how to provide the small food businesses academic training. These literatures poorly focused on-line ordering platform, so that is a new field for our study. It is significant and meaningful for us to do this exploration.

3 Methodology

We carried out survey studies and conducted interview and observation. The survey has been sent out on the internet to explore the reasons why these customers use these applications. Next we conducted interviews to investigate the complaints from customers, the cooperation of the merchants and the food security department, and merchants' concerns from the platform company's perspectives. Then observation was carried out in order to better design the applications for these companies.

This paper aims to provide design guidelines for these ordering platform companies to improve their application credibility and gain trust from their customers.

3.1 Survey

Participants. Participants were 155 Wechat users, who received the survey we sent out through group chat and Wechat moments. The sample consists of 104 students (67.1%), 44 office workers (28.39%), and 7 others (4.52%) ranging between 11 and 53 years old, with a mean and median of 32.

Methods. We carried out survey studies to group chats and moments on Wechat, since Wechat is one of the most convenient social platforms in China. And the application which was used named WenJuan xing (Translation: Questionnaire Star) to help to design the survey, since it is a convenient questionnaire software providing many models.

The survey was designed by two parts: multiple choice questions and fill in blanks. It aimed to collect the main factors which effect people most when they decided to choose one or many applications to order take-out food, and effective suggestions for the service improving of applications in the future.

3.2 Interview

Participants. Interview part focused on the applications' users, customers and applications' cooperator, the restaurants. There were three groups of interviewees in the interview. First was the staff who were responsible for customers order management and customers' complaints handling at the Meituan application, one of the most popular online meal ordering platform company. Second, we interviewed eight consumers. Four staff were interviewed from each restaurant. There are four picked restaurants: Participate One is from one of the Chinese traditional restaurants, Participate Two is from another Chinese traditional restaurant, Participate Three is from a Korean restaurant, and Participate Four is from a snack restaurant.

Methods. Phone interview was the way to compile the survey of the staff who were responsible for customers order management and customers complaints handling at the Meituan application. The face-to-face interview was for the merchants of the chosen restaurants who had cooperation with these platforms and consumers who were using online meal ordering applications. The four chosen restaurants are which cook different types of food: two of them cook Chinese traditional food, one restaurant cooks Korean food, the last one cooks some snacks including fried chicken and drinks. All the restaurants are located on the street in front of the Southern gate of one university.

3.3 Observation

Participants. The five merchants were chosen, which were on the street in front of the Southern gate of the university. The time we did the observation is lunch and dinnertime. We found that our observation time of the dinning peak was much more provable than other period. We observe two customers about their interaction with the applications.

Methods. Most small merchants set their kitchens in the same room which they serve the customers to have the meal, dividing by a reception desk. That's easy for us by chance to observe the environment of their kitchens and cooks, and we also got into their kitchen to observe the materials and their kitchenware. However, we found that there still was a merchant just have a kitchen without a dining room. In these restaurants, it is easy to observe the staffs using these applications when they were working. The whole applications' using process was observed while the two objects were using these applications as usual.

4 Result

4.1 Survey

The purpose of the survey is to find a guidance to enlighten platform to promote. Based on the collected data, we found that the target groups of take-out food applications, the most popular application which can be taken as a successful example to use for improve company's propaganda model, and the factors effecting users' choices (Table 1).

There were 155 surveys sent out. Because of Wechat survey, we received 100% response. Wechat is one of the most popular Internet social networking applications in China. As soon as people click in it, they have to finish all questions and then they will be able to submit the result. People who have answered the survey can be divided into 4

Table 1. Demographics

Age area	Population	Occupation	Population	Category	Population
Under 18	**6**	Student	**104**	Meituan	**108**
18–22	**99**	Worker	**44**	Eleme	**54**
23–35	**11**	Others	**7**	Baidunuomi	**15**
36–53	**39**			Others	**23**

Table 2. The reason why customers choose a meal ordering application (More than one choice)

Reason	Population (155)	Percentage
Pretty application interface	20	12.9%
Many kinds of restaurants can be chosen	67	43.23%
Low deliver fee	54	34.84%
Short deliver time	30	19.35%
Many price activity	43	27.74%
Others	39	25.16%

groups, under 18, 18–22 years old, 23–35 years old, and 36–53 years old. We collected the data and disposed them into a demographic table, which contains age area, occupation, category, that which kind of application people usually use.

The age data shows that 18–22 years old and 36–53 years old groups take a large part of people in the 155 surveys, 63.87% and 25.16%. The groups, under 18 and 23–35 years old, occupy only 3.87% and 7.10% of this survey.

As expected, in the persons who answered the surveys, students are the mainstream of using on-line meal ordering platforms, having 70% percentages. And, worker part take less percentages, about 29%. So, the major customers who these platform companies are facing are student.

The category of the application, which contains three major ordering applications. According to chart, Meituan has the most usages among meal ordering applications, 69.68%, more than half. Eleme takes the second place, 34.84%, less than 50%. Baidunuomi takes the last place, 9.68%. To our surprise, there are other ordering applications taking 14.84%.

Table 2 shows the reason why customers choose a meal ordering application. Through the chart, I could know that customers pay 43.23% attention on “Many kinds of restaurants can be chosen”, comparing to other reasons, closing to 50%. The second one is “Low deliver fee”, occupying 34.84%. And, “Pretty applications interface” takes 12.9% in this survey, but there are still some people concern about that.

Figure 1 deals with the way knowing ordering applications. Most of people know these applications through their friends’ recommending, having 52.26%. Second efficient expression way is “App store recommending”. The useless way is “Paper advertisement”.

According to the advice bar chart, we found that customers do concern about these aspects: sanitation, deliver time, price activity, better organization, more food choices,

Fig. 1. Advice from customers

protect customers privacy, food quality and service. Many people think about sanitation, deliver time, price activity and organization these four parts. And there is a large proportion of people are satisfied with present applications' usage.

In summary, the survey shows that young people are the mainstream in platform using team, which means companies should make advertisement mostly focusing on these age groups. The result shows that these young people are mainly students and office workers. Therefore, whenever company plan to popularize its application, such as advertising GIF banners on Internet and passing out leaflets, they should make a strategies that introduce factors attracting young people mostly. Furthermore, company can also take these young people as target group to do survey to collect useful feedbacks.

Secondly, the results shows that Meituan is the most popular application, so its development style can be an example for other companies to make a case study, which will help companies to maintain its popularity and reputation.

Thirdly, customers concern about not only an application itself development, but also the corporation with merchants, involving restaurant choices, price setting and deliver time. According to expression way part we found many people received the app stores' recommendation and website advertisement besides friends' recommendation. In that case, providing more awards to promote users to share the application would be a smart choice.

Even though we got a high survey response rate, there are still some limitations of our survey. First, the majority of the participants in our Wechat surveys are our friends who are also usually students and some parents. That's the reason why 18–23 and

36–52 years old groups have a large number of percentage and it may influence the final result. There are more adults in the society have not been investigated. Finding an open way to deliver surveys is next step we can explore for the survey.

4.2 Interview

The interview is to investigate the customers' using situation and then provide an evaluation for a popular platform, in order to find some correlation among online meal ordering platforms, merchants, and consumers.

Staff of the Application Company. Through the interview, we could gain that 10.3% people thought their satisfaction level was 1 (dissatisfaction), 18.2% people belong to 2 level, 28.3% people belong to 3 level, 23.1% people belong to 4 level, 20.1% people were belonging to 5 level (very satisfied). And for the reasons of complaints, most people are dissatisfied with the quality of the meal, such as the food isn't fresh, or the matching rate, that foods look a little bit different from pictures shown online. Next major complaint reason is delivery service, for example, after ordering, the customers waited for their meal for a long time. Except other options, there are a few of people are concerning about the usage of applications as follows.

And from the data, we could know that most people are satisfied. But people have not always the same taste, so there are still some people complained about the meal. (Participate A)

Staffs of the Restaurants. The data from this interview part could provide the feedback of the consumers from corporation restaurants side. Three staffs expressed that they like this kind of applications, which is quite convince for their daily life. They are all satisfied with this application, and its using process is quite convenient, although different people still found there are few parts which can be proved.

I like this application what our restaurant use. It increases my job speed. And I will post some questions which I met when I use the application, and when I check it next time, it usually has already improved. (Staff A)

It (application) is quite good and very easy to learn. I think my work can not be function without it. But there are some customers are really picky that they sometime write down the negative comments unreasonable for our business which did influence our business. (Staff C)

The application helps me a lot. In the pass, we did take-out food by telephone. If there was something wrong with the telephone net, missing some information, it would have leaded to plenty of troubles. At present, using this kind of application can avoid this situation and we can work faster than before, which does satisfied by customers. (Staff D)

And one staff is not satisfied with application, due to the price reason.

I found that the price the customers paid occasionally is not the data which we posted. I have no idea about that if the customers do not complained about the price. I thought maybe the application company does adjust, but they did not tell us. (Staff B)

Customers. By consumers' part, almost all consumers expressed that these applications are convenient, having become a part of their life. And it would be hard for them to live without these applications. After the interviews, interviewees gave these platforms their opinions:

75% participants ask for the guarantee of the environment of the restaurant and their deliver time. 50% of participants hope that the applications could provide more price activity. And all participants expressed that these platforms had better provide a better organization of the page frame, including less useless advertisement. They did either put their attention on the food quality and service quality.

The purpose of the interview to is to investigate the how users think about these platforms and their opinions or complaints, in order to find what the challenge of on-line meal ordering platforms may face. Based on the staff who responsible for customer order management we interviewed, we found most people chose to give proper judgements, that there is no one-side judgement. And most people are not satisfied due to the quality of the meal. Consequently, the company did make some adjustment according to customers' advice or complaints. Although, at present a large part of people care more about the food quality or deliver service, there are still some consumers do concern about the applications' using.

After analyzing the feedback from staffs, we could get that most of them have already got used to use this kind of internet tool to increase their business. Thanks to these platforms, they could gain consumers feedback. And if consumers had meal in their restaurants, it would be more difficult to gain their opinions. But when they use these platforms, it is not always favoring, because there are still some picky people who usually will disturb their business and gave their negative comments without any real reasons. These comments are useless for their work. Fortunately, major comments are useful and can help them to improve their work.

According to the interview, we could find that the price always can be the most important thing except the problem of food quality. Because some restaurants require the basic price which customers need to pay, and if the price customers have ordered matches or is higher than the basic price, these merchants will provide deliver service. And usually this basic price is higher than the average price which people will pay while going to the restaurants. That's the reason why usually customers want many discount for their goods and service.

And when consumers use the applications, they are really in hurry. They expressed that they do want to find what they want directly and quickly. That's why they prefer to seeing the order page easily and clearly. So, customers suggested in the applications that they hope company designers can reduce or better organize the advertisement. In the same time, food quality and service quality are two significant parts which not want to be ignored by customers.

The result of interview shows that although a part of people are satisfied with present platforms, there are still some parts which platforms' companies can work on, such as price activities, page frame design. Through the interviews, most of the customers express that they would like to see more price activities when they order. So the price problem might be a direction for the companies and the merchants to concern about, besides improve the service. Another is about the page frame of the applications. In the interview, interviewees complaint that the home page of the platforms include useless information, and it's not convenient for them to find what they want quickly or explore new restaurants. So the page frame and information arrangement have many potential to be developed. Unfortunately, in my investigation, we have not concerned about the condition between eat-in and carry-out and analyze the reason why people choose to use online meal ordering instead of going to the restaurants, or on the contrary. This is an area that we can develop in the future. And we have only concerned about the opinions from merchants to the platforms and companies, and ignored the suggestions from the other side, because cooperation is from both sides. If there is opportunity, we could also interview the opinion about the merchants from the platforms in order to gain more information.

4.3 Observation

According to the observation, we found that restaurants observed can be classified into two parts: the one is only providing take-away business which contains only kitchen in their shop, the other is not only do take-away business, but also service the customers directly in their store. The former's kitchen are not very clean and tidy compare to the latter, especially the ground in front of the store. They just have a kitchen for cooking things, and their food materials does not look so fresh. After cooking, the meal will be packed well by beautiful boxes which have passed the food safety inspection, and orderly arranged to different places. Different from the former, the latter restaurants are usually use materials which are fresh and healthy, and their kitchen wares seems cleaner. Also, their decorations are delicate and beautiful, because they provide service that consumers can have meal in their shop. Clean and ordered tables make people feel much safer and satisfied. All cooks looks are very nice, same clothing with white apron and tall cook hat, making people feel comfortable and believe in them. Through this observation, these restaurant we have been mostly match the same as the pictures which they published on the software.

By using process, we observed two people when they were using the applications. And we have seen how they click the applications, the process they order their meals and how they used these online platforms to search the food they would like to eat. They both have already mastered the usage and they said it is very easy for new user to handle them, but there is only one thing that there are too many extra function areas which may disturb people at the beginning.

By the way, during the observation, we found that merchants usually choose 'Eleme' and 'Meituan' to develop their take-away business, but there is an interesting situation, that sometimes people selected their food on 'Eleme' or other apps, but when they received their food, they found that the deliver worker wear other company's clothes, making them confused. This phenomenon may be explained in three potential reasons:

The first one is that the platforms' companies do not organize well; the second is that deliver workers are not enough, especially at peak time, so the merchants share the deliver worker with each other; the third is about the deliver workers' part-time job—maybe these workers are working for different company at same time in order to get more salary, so they can just wear one company's uniform.

Actually, some customers have either found this phenomenon and they are either confused about that. This answer can also be gotten from interview part. To be honest, what they care about is not about the uniform which deliver workers wear, but the food quality and service which company provide and whether it is good or not. And confused and unlatched deliver company may cause their worried and trust problems. So, we suggest that the related departments can reinforce the management, and formulate some policies.

After observing two people using meal ordering platforms, we found that there are real some useless and disturbing advertisement on the platforms' pages and they are real a trouble for users to use, which approved the opinions from interview part. Frankly speaking, most ordering process is easy to be mastered, but there are some information of the pages are not so expressive, which may affect the user experience. And according to the survey part, most user are young and middle ages and the elder age is missing. In our opinion, the using step of platforms is easy, but still not so easy for the old to have try. In modern China, plenty of people start using internet technology, and the old are either joining us. So, platforms companies can do some try in this part.

Honestly speaking, this observation may be not so perfect. Because of our identity and conditions limitation, we could just observe merchants around our school. In later work we'll try to choose more merchants or merchants at different places to perfect my observation, if it is possible in the future to continue.

Our academic target of this paper is to help these ordering platforms companies and provide design guidelines for them to improve their design and credibility. It seems that we get a pretty good result, however that there are several parts we need to improve in the future.

5 Conclusion

Due to social event about lots of small-scale manufacturers without business licenses sell their food though the on-line platforms, and this situation has been questioned by customers about the credibility of these applications and caused trust issues, our group chose this topic to do the investigation. We investigated the challenges that these on-line meal ordering platforms are facing, for example, website design, customers' trust and its application credibility. The results of our investigation from three methodologies do support each other. Observation part supports the survey part data, that most of people are keen on Meituan and Eleme applications.

This paper aims to provide design guidelines for these ordering platform companies to improve their application design and credibility, and gain trust from their customers. After a series of investigation, I gained a lot of meaningful information and achieved my purpose that our result can give a specific guidance for meal ordering platforms. However, we still have some parts which can be improved in the future. If it is possible, I will continue working on specific parts about how on-line meal ordering platforms can follow our advices to improve their work.

References

1. Fu, B., Lin, J., Li, L., Faloutsos, C., Hong, J., Sadeh, N.: Why people hate your app: making the sense of user feedback in a mobile app store. In: Proceedings of the 19th ACM SIGKDD International Conference on Knowledge Discovery and Data Mining, pp. 1276–1284 (2013)
2. Yin, P., Luo, P., Lee, W.-C., Wang, M.: App recommendation: a contest between satisfaction and temptation. In: Proceedings of the 6th ACM International Conference on Web Search and Data Mining, pp. 395–404 (2013)
3. Liu, S.M.: The use of mobile and wireless technology and food quality control in China. In: Proceedings of the 6th International Conference on Theory and Practice of Electronic Governance, pp. 518–519 (2012)
4. Gupta, R.: Food security and safety using advanced information and communication technologies (ICTs). In: Proceedings of the Second International Conference on Information and Communication Technology for Competitive Strategies, p. 58 (2016)
5. Gerson, R.: Measuring Customer Satisfaction. Crisp Learning (1993)
6. Swift, T.: Trust, reputation and corporate accountability to stakeholders. Bus. Ethics A Eur. Rev. **10**(1), 16–26 (2000)
7. Chen, C., Chen, P., Huang, C.: Brands and consumer behavior. Soc. Behav. Pers.: Int. J. **40** (1), 105 (2012)
8. Boyce, J. Broz, C.C., Binkley, M.: Consumer perspectives: take-out packaging and food safety. **110** (8) (2008)
9. Egana, M.B., Raats, M.M., Grubb, S.M., Eves, A., Lumbers, M.L., Dean, M.S., Adams, M. R.: A review of food safety and food hygiene training studies in the commercial sector. **18**(10) (2007)

An Investigation of Weak Ontology for New Media Studies

Edward Chor Kin Hung(✉)

Department of Journalism and Communication, Faculty of Arts,
Hong Kong Shue Yan University,
10 Wai Tsui Crescent, Braemar Hill Road, North Point, Hong Kong
ckhung@hksyu.edu

Abstract. Recent studies reveal the importance of ontology in new media studies. However, there is not yet a commonly agreed new media ontology for media studies researchers and practitioners. Some scholars regard this insufficiency as the impracticality of localization in the existing ontologies concerning new media. We contend that a weak ontology that allows localization to happen should be carefully examined when developing a new media ontology. This paper is to disclose this potential of weak ontology to the new media community.

Keywords: New media · Weak ontology · The identity of indiscernibles

1 Introduction

Recently there have been studies about the importance of ontology in new media. Srinivasan [1] claims that the ontologies for new media allow media studies researchers and practitioners to better utilize the potential of the digital world surrounding us. An example is Web 2.0 that allows users to "create and share metadata according to their own local experiences and interpretations" [1], but is limited to websites. Web 2.0 implies the commonalities shared by different communities. Srinivasan also suggests fluid ontologies, which gather information from users, to be the ontologies for new media. However, fluid ontologies ignore generalizations, focusing on domains. Kittler [2] emphasizes that forming an ontology for new media is "both a crucial and a difficult one" [2] that requires the ontology to know "how to read blueprints, layouts, mainboard designs, industrial roadmaps, and so on" [2] especially when it is supposed to be technically informed and domain-specific. In short, there is not yet a commonly agreed ontology designed for new media. Thus, our question is how we should develop a new media ontology. This brings in two sub-questions: Where should we start? What is the reference?

2 Methodology

In this paper, we adopt content analysis as the research method. We firstly examine the literature about the possibility of having an ontology for new media. Then, we review the previous works regarding the development of a new media ontology and the

A.G. Ho (ed.), *Advances in Communication of Design*, Advances in Intelligent Systems and Computing 609, DOI 10.1007/978-3-319-60477-0_14

insufficiencies. Lastly, we explore the potential of a weak ontology in the development of a new media ontology.

3 An Overview on the Recent Endeavors to Develop a New Media Ontology

3.1 The Possibility of an Ontology for New Media

There are scholars who find having an ontology of communication impossible. Kallinikos et al. [3] claim that digital artifacts in the form of new media possess an ambivalent ontology and further refer to the findings of Ekbia [4] that such an ontology violates Leibnitz's law of the identity of indiscernibles:

The identity of indiscernibles represents an ontological property in which if x and y have the same properties, x is identical to y.

Their standpoints are that since the digital artifacts from new media are "fluid and editable, often embedded in complex, distributed, and shifting digital environments" [3], it is very difficult for them to share the same set of properties and be identical, thereby violating the identity of indiscernibles.

We argue that their findings only represent a particular conceptual level of these digital artifacts. If we further generalize these digital artifacts, we find that they do share the same set of ontological properties at a more abstract level, maintaining the identity of indiscernibles. We will further provide information on this matter in the section on weak ontology.

3.2 The Previous Works on the Development of a New Media Ontology

Srinivasan [1] contends that we should consider local ontologies, especially non-Western ones when designing new media technologies. These local ontologies are capable of driving "the architectures, algorithms, and languages of new media" leading to the "social construction of technology and systems." In this way, new media technologies are built upon "multiple, diverse ontologies." Srinivasan uses the terms "ontological multiplicity" and "culturally diverse ontologies" to emphasize the importance of these non-Western local ontologies for new media technology development. He also suggests two ontologies to do so. The first one is the fluid ontology. It is "a set of classifications and semantic relationships expressed by a group of community members used to design a database in a locally appropriate manner" [1]. However, certain key opinion leaders in the community might form a bias in the fluid ontology, reducing its reliability. The second one is the process ontology for business that "links activities with one another in the context of achieving a theme or task" [1] implying the importance of experience in ontology development. In addition, Srinivasan emphasizes that a new model may be necessary to allow global new media projects to be developed with multiple ontologies. An example is the Emergent Databases-Emergent Diversity (ED2).

From our perspective, Srinivasan has not yet detailed how the fluid ontology can hold the different classifications and semantic relationships together. Also, he does not pinpoint what criteria we should use to classify our experiences in the process ontology. Likewise, he does not explain how a new model can hold the differences among the different communities together and how a piece of information can be understood among them efficiently. We regard all these issues as lacking the underlying ontological properties shared by the fluid ontology, process ontology, and any relevant ontologies targeting ontological multiplicity and culturally diverse ontologies. Under these circumstances, we propose an exploration of a weak ontology to solve these issues.

4 Weak Ontology

According to Hung and Choy [5], a weak ontology allows different domain ontologies to be developed from it. In other words, a weak ontology defines the ontological properties shared among different domain ontologies. A weak ontology is more conceptual or abstract than the relevant domain ontologies. An example is their creative work ontology for cultural product creation. They claim that such an ontology has structures and features as its ontological properties, rules as its axioms, and biases as the functions of the features specifying the manifestation of a cultural product. Since the creative work ontology is a weak ontology, different domain ontologies representing different cultural products including popular music, clothing fashion, and narrative can be developed from it. Figure 1 shows a UML diagram of the creative work ontology. Figure 2 shows a domain ontology, namely, 2D shape ontology developed from the creative work ontology.

Another example is the electric guitar ontology by Bandini et al. [6] as shown in Fig. 3. It is meant for the ontological development of different electric guitars.

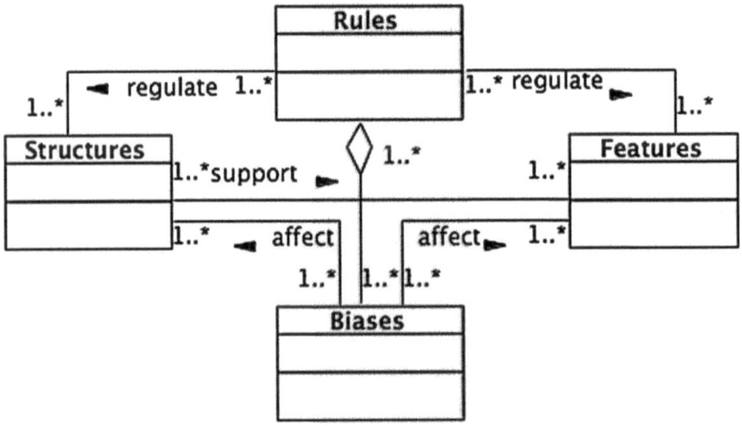

Fig. 1. A UML diagram of the creative work ontology – a weak ontology – that allows the development of the relevant domain ontologies (as cited in [5]).

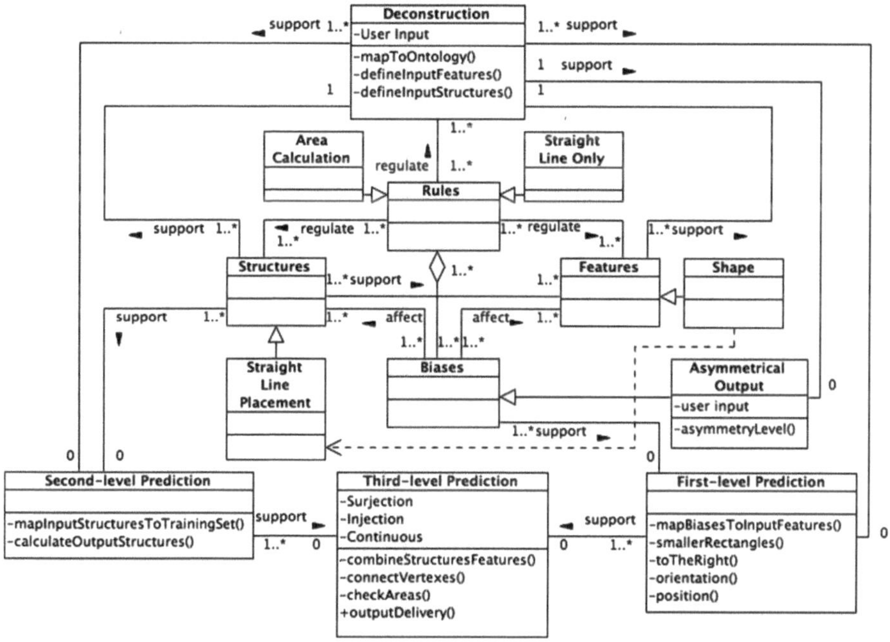

Fig. 2. A subset of the 2D shape ontology developed from the creative work ontology. This UML shows the working relations between the subset and conceptual recombination, a creative method designed for creative systems (as cited in [5]).

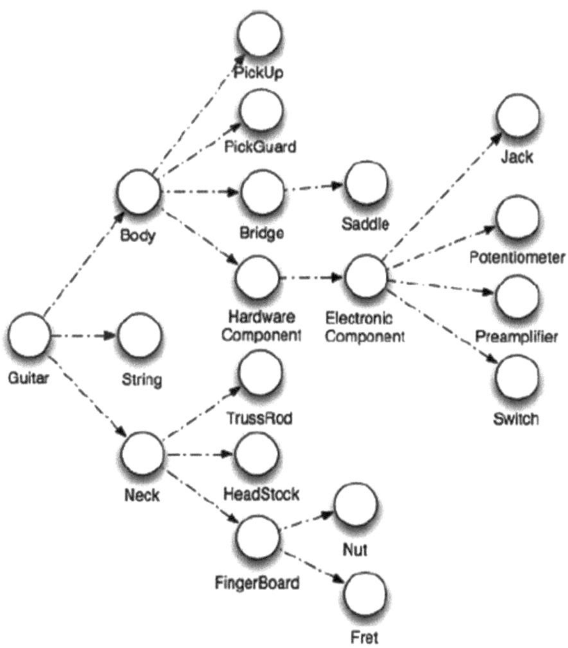

Fig. 3. The electric guitar ontology with the circles as the classes (as cited in [6]).

In the respect of developing a new media ontology, we should consider adopting a weak ontology so that it can "resonate with cultural and community diversity" and then direct other domain ontologies to develop based on "their diverse social, cultural, and political goals" [1].

5 Discussion

In this paper, we pinpoint the need for a weak ontology for the development of a new media ontology. This is due to the fact that the missing commonly agreed new media ontology in the past can be formed at a more conceptual level in which the relevant domain ontologies of different new media can share its ontological properties and develop from it, giving rise to ontological multiplicity and the culturally diverse ontologies of new media. Brummett and Duncan [7] state that, "critical studies typically [ground] the components of the process of communication in a social, discursive ontology" [7]. They further claim that, "To identify a shared discursive ontology underlying the categories of communication is to see them as created together, simultaneously, through the same discursive processes" [7]. Though we are not claiming that the new media ontology is a discursive ontology, it is possible to suggest that Brummett and Duncan's "created together, simultaneously" can be interpreted as sharing the same set of ontological properties among these categories regardless of the timestamps they are attached to. In this way, a weak ontology deserves our serious consideration for the development of the new media ontology.

Moreover, we claim that the digital artifacts of new media are indeed the representations of new media and share the ontological properties of cultural products. According to the identity of indiscernibles, cultural products and the digital artifacts are hence similar to each other as they share the same set of ontological properties. Likewise, the digital artifacts are also akin to each other when they share the same set of ontological properties. The ontology of the digital artifacts is no longer ambivalent.

On the other hand, there is creative work ontology – a weak ontology – that describes the properties of cultural products. In regard to the intimacy between cultural products and digital artifacts, it is feasible to use the creative work ontology as a foundation to develop our proposed new media ontology.

In the future, we will develop a new media ontology in view of the findings of this paper.

References

1. Srinivasan, R.: Re-thinking the cultural codes of new media: the question concerning ontology. New Media Soc. **15**(2), 203–223 (2013)
2. Kittler, F.: Towards an ontology of media. Theory Cult. Soc. **26**(2–3), 23–31 (2009)
3. Kallinikos, J., Aaltonen, A., Marton, A.: The ambivalent ontology of digital artifacts. MIS Q. **37**(2), 357–370 (2013)
4. Ekbia, H.R.: Digital artifacts as quasi-objects: qualification, mediation, and materiality. J. Am. Soc. Inf. Sci. Technol. **60**(12), 2554–2566 (2009)

5. Hung, E.C.K., Choy, C.S.T.: Conceptual recombination: a method for producing exploratory and transformational creativity in creative works. Knowl.-Based Syst. **53**, 1–12 (2013)
6. Bandini, S., Bonomi, A., Sartori, F.: GUITAR HERO, a knowledge-based system to support creativity in the design and manufacturing of electric guitars. In: Gero, J.S., Goel, A.K. (eds.) Design Computing and Cognition 2008, pp. 181–200. Springer, Heidelberg (2008)
7. Brummett, B., Duncan, M.C.: Toward a discursive ontology of media. Crit. Stud. Media Commun. **9**(3), 229–249 (1992)

Consistency of Visual Information in Web Design - Focusing on Responsiveness of a University Website

Toshiki Matsuo[1(✉)], Wonseok Yang[2], and Naoya Shibata[1]

[1] Shibaura Institute of Technology Graduate School of Engineering and Science, 3-7-5, Toyosu, Koto-ku, Tokyo, Japan
md16079@shibaura-it.ac.jp

[2] Shibaura Institute of Technology Engineering and Design, 3-7-5, Toyosu, Koto-ku, Tokyo, Japan

Abstract. Most responsive websites are not very consistent. Thus, we revealed the elements and the factors that enable consistency in responsive websites in a multidevice context. We investigated the available research and the present state of university websites. We performed an evaluation experiment with five responsive websites and created 15 sample sites. We did an evaluation experiment with this. Consequently, we were able to obtain the elements and factors. We think that we can achieve a more natural multidevice experience through this knowledge.

Keywords: Responsive design · Consistency · Multi device · Visual information

1 Introduction

Web design based on the PC has developed significantly and has provided much information for users. However, the surroundings in which people use the web were made complex by the arrival of other devices. Thus, communicating the most suitable information to a Web user using conventional web-design conditions has become difficult. In addition, the quantity of information has been increased through realistic expression styles like skeuomorphism. To overcome this issue, flat design and responsive web design were implemented. Responsive web design is a technique based on CSS of HTML5 and can change the layout of the website depending on the screen size of the device [1]. However, it is difficult to ensure the same user experience for all device users because many responsive web sites only change the layout, font, etc. depending on the screen.

In recent years, through flat design, such as the UI design of Windows 8, the visual spread and amount of information has become "easy to see," "easy to know," and "easy to understand." Also, the UI design of iPhone from iOS 7 onwards changed significantly to something like a flat design, and usability became better. Given these trends, we think that users can have a better experience by reconsidering not only surface

A.G. Ho (ed.), *Advances in Communication of Design*, Advances in Intelligent Systems and Computing 609, DOI 10.1007/978-3-319-60477-0_15

design but also the relationship between human and device for a particular content or information.

In future, in the age of the Internet of Things, the device connection network is expected to increase. Therefore, we believe that we need to create guidelines for designing a responsive website for the same user experience in a multidevice environment.

In this study, we abstract the elements relating the consistency of the responsive website on various devices and clarify the influences of these on user experience. We propose guidelines to design a responsive website such that users may get the same experience with these elements.

2 Research on Japanese University Websites

2.1 Survey Content and Purpose

Previous studies show that domestic universities are responding to the rapid popularization of multidevice usage [2].

We studied whether each of the abovementioned websites is responsive (05.25.2015–06.17.2015). The contents of this research were as follows:

- Whether or not the website is responsive
- Increase or decrease in the content
- Type of menu

2.2 Survey Method

We browsed a given website with three devices—a desktop PC, a tablet PC, and a smartphone. The specifications of each were as follows (Fig. 1):

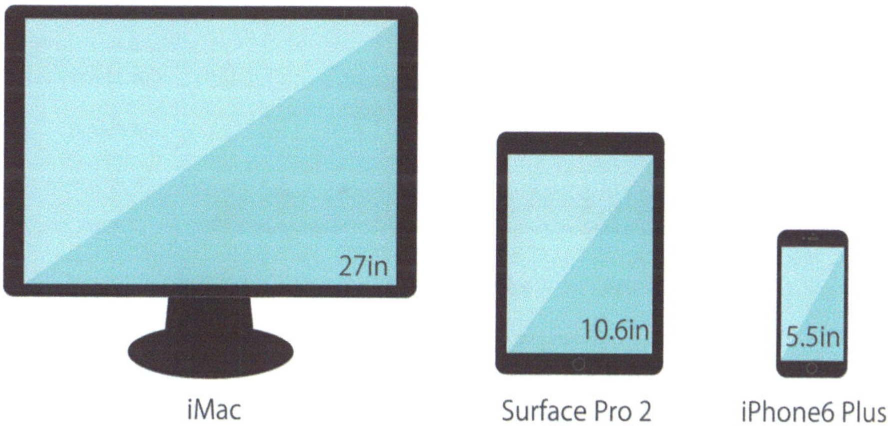

Fig. 1. Screen size ratio of the device used for survey

- Desktop PC: iMac (Size 27 in; Resolution 2560 × 1440)
- Tablet PC: Surface Pro 2 (Size: 10.6 in; Resolution: 1920 × 1080※)
- Smartphone: iPhone 6 Plus (Size: 5.5 in; Resolution: 1920 × 1080)
- ※Surface Pro 2 is used in portrait orientation.

Navigation patterns were classified according to the five patterns introduced in [3].

Figure 2 shows the ratio of responsive websites. In national universities, the websites were slow in keeping up with multidevice requirements, but 30 per cent of the websites keeping up with multidevice use were responsive websites. In nongovernmental universities, nearly 70% of the websites kept up with multidevice requirements, but many of those were to be accessed exclusively on smartphones.

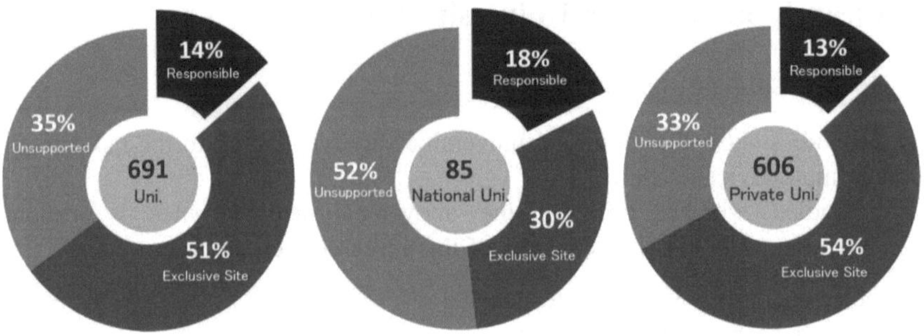

Fig. 2. Screen size ratio of the device used for survey

Figure 3 shows the increase or decrease in content. A total of 30% of the websites for exclusive access through smartphones have different content on each device. In addition, many of these websites decreased the content. However, most responsive websites did not increase or decrease the content and provided access to the same information.

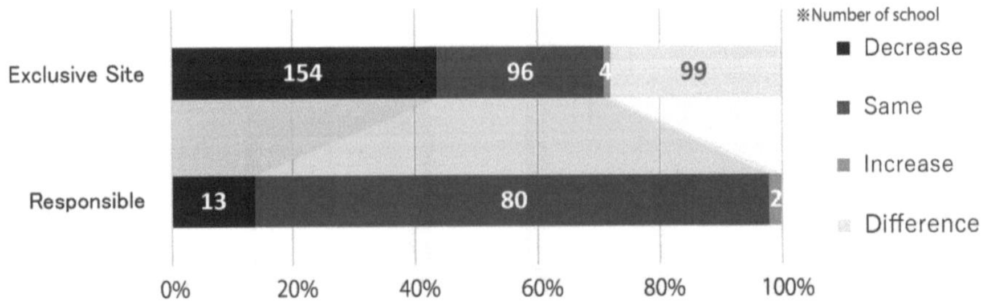

Fig. 3. Increase or decrease of content by using responsive web design

According to this survey, we found that responsive web design is not used much in domestic college sites. However, responsive web design is a new technology, and its use was higher than expected. Because of Toyama University's introduction of responsive web design in fiscal year 2015, there has been a major improvement in usability. From this, we believe that the number of university sites using responsive web design will increase.

Also, in sites using responsive web design, common content was easy to handle, so there was hardly any increase or decrease in content between devices. For this reason as well, responsive web design is an effective method for creating a multidevice-compliant website.

3 Abstracting Elements that Relate to Consistency and the Factors Arising from the Same

3.1 Plan of the Experiment

We abstracted eight elements—namely, "color," "font," "text of content," "fluctuation of contents," "image of contents," "first view," "layout," and "menu"—from our research on responsive websites of Japanese universities and some literature [4, 5]. Moreover, because we drew out the factors that relate to consistency from among these elements, we had 20 university students do an evaluation experiment with the following order (10.15.2015–10.17.2015).

Experiment Conditions. The university websites used for this experiment:

We used five university sites that can allow comparison of the factors, and use responsive web design.

Toyama University, Kitami Institute of Technology, Kansai University of Health Sciences, Dohto University, and Gakushuin University

The devices used in this experiment were:

- Desktop PC: iMac (Size: 27 in; Resolution: 2560 × 1440)
- Smartphone: iPhone 6 Plus (Size: 5.5 in; Resolution: 1920 × 1080)

Flow.

1. Browse and compare each website on the desktop PC and smartphone.
2. Evaluate the eight factors, on a scale of five, on how consistent they feel on each site.
3. Subject writes the reason for their evaluation of each element.

3.2 Result

Through this experiment, we obtained the factors as given in Table 1. In this experiment, since we evaluated each item by looking at the entire site, we were able to find various factors for each item.

As a result, we obtained many factors. Through this experiment, we were able to discover many factors for every element because our subjects evaluated each element while assessing each website in its entirety. However, it is difficult to determine

Table 1. Factors considered part of consistent experience

Elements	What is considered to be a factor of consistency
color	・Align the colors of menu bars and headers. ・Use the theme color as the main. ・Do not use lots of colors. ・Adjust color sense of volume.
font	・If it is Gothic, there is sense of unity even if the font is different. ・Adjust the size feeling by screen size (range of readable size). ・Differences between letters do not compromise much consistency.
fluctuation of contents	・Number and amount of information are the same. However, if the contents are large, the screen becomes longer as the screen becomes smaller, and the consistency is lost. ・Simplification and storage can maintain consistency (length consistency).
text of contents	・It is more consistent if the amount of text does not change. ・Consistency can not be felt when lines change or size ratios are different.
image of contents	・Unify the image and color of the content. ・Keep the size of the image. ・Even changing the aspect ratio can maintain consistency (however, color and image are unified).
first view	・Align the header of the site. ・Adjust logo and key visual. ・Unify the impressions of the site (how to look).
layout	・Match the flow of your eyes. ・If changing the layout with the amount of content left unchanged, consistency is difficult to obtain because the length becomes long.
menu	・Block type navigation patterns are easy to achieve consistency. However, if the item is an odd number, blanks will be created, and measures will be required to lose consistency. ・If color, position and contents are the same, consistency can be maintained even if it is stored.

whether the factors obtained in this experiment are relevant for obtaining a truly consistent experience. Thus, we need to conduct an experiment to prove that these factors are correct.

4 Experiment to Clarify Elements and Factors that Influence Consistency

4.1 Plan of the Experiment

We created 15 sample websites and applied the factors obtained above to the website of "Shibaura Institute of Technology (11.13.2015)." Because we mentioned that these factors provided consistency in a multidevice environment, we had 15 university students perform an evaluation experiment with these samples (12.21.2015–12.30.2015).

Flow. Our subjects did the following for each sample:

1. Browse and compare the sample website on the desktop PC and smartphone.
2. For the sample website, evaluate four items on a scale of seven.
3. Compared to the relationship between devices on the base site, evaluate on a scale of seven the consistency of the site with the modified feel.

Evaluation Item. Evaluate each in seven steps.

- How consistent does the modified site feel, as compared to the relationship between the devices for the base site.

The following items are evaluated for the base site and the modified site:

- Overall visibility
- Understandability of information
- Readability of letters
- Consistency of design within the devices

4.2 Plan of the Experiment

The results of the evaluation were mapped in two patterns, as shown in Figs. 4 and 5. Figure 4 is mapped based on the evaluation of usability and the evaluation of consistency. Figure 5 is mapped based on the evaluation of consistency in appearance and the evaluation of consistency in use obtained from evaluation on usability.

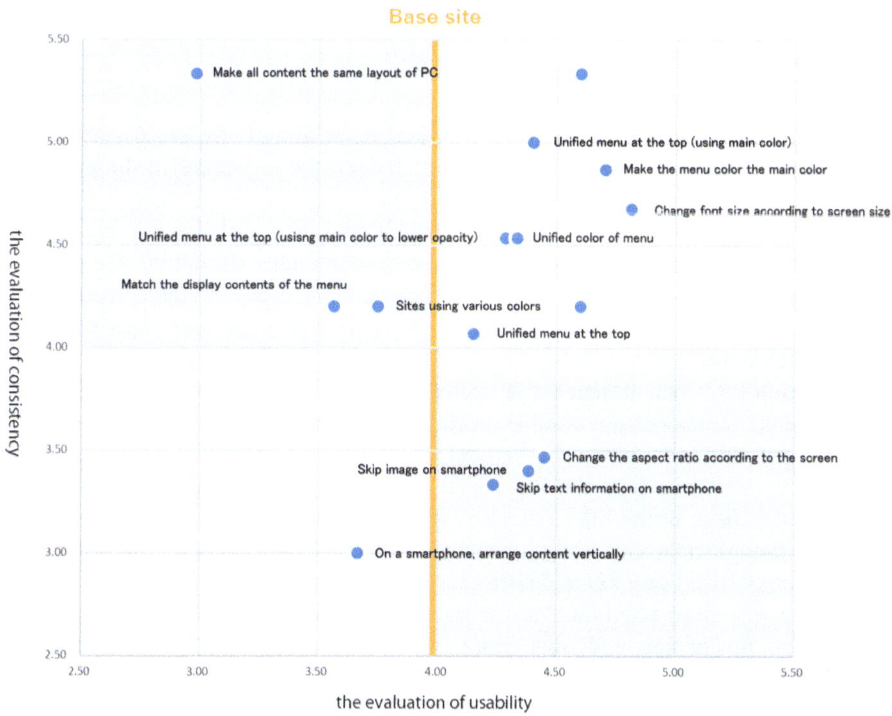

Fig. 4. Map based on the evaluation of usability and the evaluation of consistency

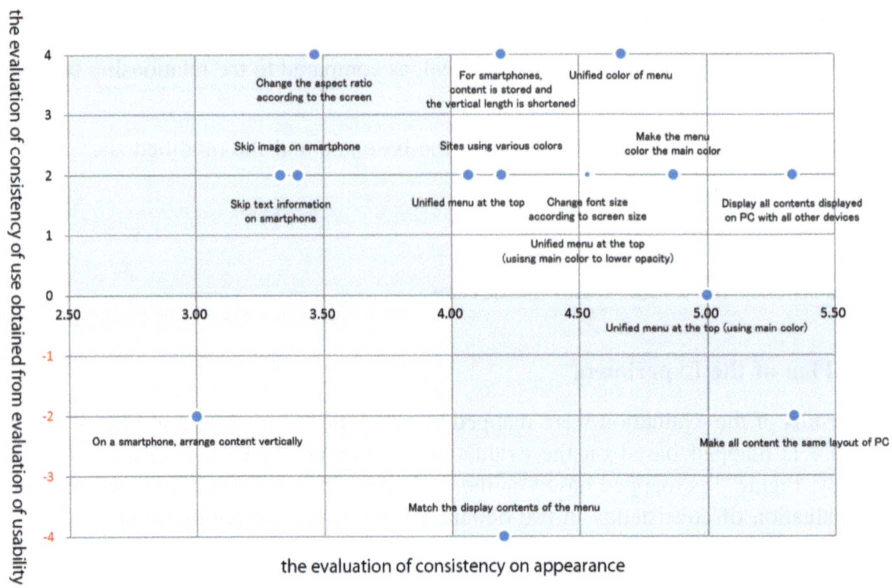

Fig. 5. Map based on the evaluation of consistency on appearance and the evaluation of consistency of use obtained from evaluation of usability

From these mappings we abstracted the elements relating to the consistency of the responsive website on multidevice usage and clarified the influences of these on user experiences.

From this result, Fig. 6 shows the elements for consistency of a responsive website for various devices and factors to obtain the same experience in multidevice usage.

In this experiment, we were able to abstract elements related to the consistency of responsive websites on various devices and clarify their effect on user experience with quantified evidence.

The result for "content increase or decrease" was surprising. Contradictory designs that "make the same content viewable on all devices" and "store the contents according to the screen size and shorten the vertical length" were factors that helped maintain consistency. Also, matching the PC design to the design of the smartphone, such as "unifying the menu at the top" was also effective for maintaining consistency. From this, it is thought that consistency can be maintained easily when responsive web design is done based on mobile devices such as smartphones.

Also, it was unexpected that "to use many colors" became a factor in maintaining consistency. In keeping with this result, some of the opinions obtained from the experiments were, "it is easy to understand the image because the color is determined by place" and "it is easy to understand where you are moving and where you are when you scroll".

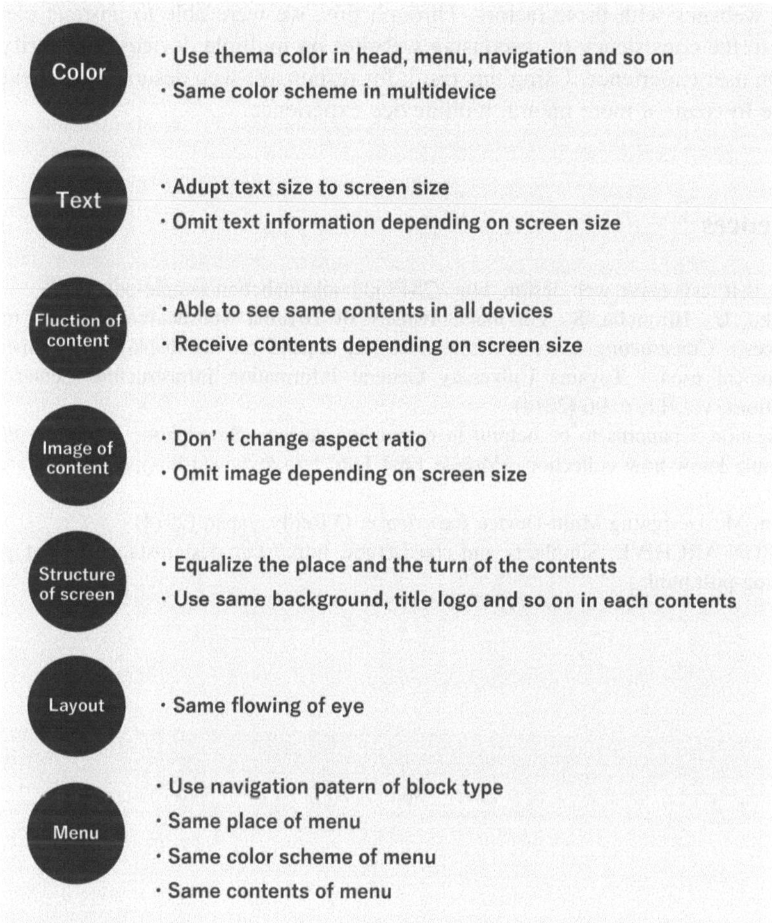

Fig. 6. Elements for consistency of a responsive website for various devices and factors to obtain the same experience in multidevice usage

5 Conclusion

In this study, we abstracted elements related to the consistency of responsive websites on a various devices and clarified their effect on user experience. We proposed guidelines for designing a responsive website that enables users to get the same experience across devices with these elements. We researched the history of media, basic rules of visual design, and the websites of Japanese universities. From these researches, we thought that "color," "text," "fluctuation of contents," "image of contents," "first view," "layout," and "menu" were elements related to the consistency of responsive websites on various devices. Furthermore, by evaluating a number of university websites for each element, it was possible to extract specific factors for obtaining a consistent experience among the chosen devices. Next, we made some sample websites with these factors. Through this, we were able to abstract elements related to the consistency of responsive websites on multiple devices and clarify their effect on user experience. Using this result for responsive web design, we believe, it is possible to create a more natural multidevice experience.

References

1. Why is it responsive web design. http://2843.jp/books/nabebon/sample-pdf-3.pdf
2. Noriko, U., Hironobu, S.: For the University of Toyama website that supports multiple devices - Constructing a website that does not depend on the display environment and equipment used - Toyama University General Information Infrastructure Center Public Relations, vol. 11, p. 96 (2014)
3. Navigation 5 patterns to be helpful in responsive design—Smartphone & mobile EC case example know-how collection—Mobile First Lab. http://www.aiship.jp/knowhow/archives/7128
4. Levin, M.: Designing Multi-Device Experience. O'Reilly, Japan (2014)
5. DESIGN ARCHIVE: Simplicity and consistency. http://stam-design-stam.blogspot.jp/2011/11/blog-post.html

Awareness Collaboration with Users and Designers: Relationship Between the Image Schema of Spoken Language and the Spatial Path of Sign Language

Zhengyu Tan[1], Shijing Fang[2(✉)], Wei Li[2], and Shihui Xu[2]

[1] State Key Laboratory of Advanced Design and Manufacturing for Vehicle Body, Hunan University, Changsha, China
Tanzy2004@126.com
[2] School of Design, Hunan University, Changsha, China
{sj_fang, liwei2014, xushihui}@hnu.edu.cn

Abstract. The main purpose of this paper is to reduce the awareness conflict between users and designers with methods of linguistic metaphor analysis and spatial gestures based on image schema. Two experiments are designed in this paper. The research methods include situation simulation, oral report, Image schema extraction and path extraction from the gesture recording. The results of these experiments are integrated to generate a user requirements interpreting system. Visualized options are provided by capturing the spatial gestures from users so that language is transformed into a visualized expression. Spatial model can be established by capturing the spatial sign language paths of the users on the basis of image schema, which provide the designers with clearer design requirements guidance and a communication platform for both users and designers.

Keywords: Perspective conflict · Spatial gestures · Image schema · Oral report

1 Introduction

The demands of users have been pointed out emphatically once and once again. However, there are still many obstacles between users and designers due to the different expressing methods and cultural background as well as the limited expressing approaches, Design is a complex process which involves perception and decision-making of different roles in different stages. The same information given by different objects will cause different actions, which may result in conflicts during the design process. Although many designers have made efforts to reduce the perceived conflicts between users and designers, it is still not possible to meet the demands of each character with different perception.

People usually use language to express their views while mapping their own first-hand experience image schema such as up - down, Internal - external, inside - outside, left – right into concrete entity. Image schema, an important concept in cognitive linguistics [1], refers to the abstraction of people's activity with the outside

A.G. Ho (ed.), *Advances in Communication of Design*, Advances in Intelligent Systems and Computing 609, DOI 10.1007/978-3-319-60477-0_16

world. Meanwhile, gestures, as an interactive input way, are now widely used in interactive design. Users can finish the harmonious man-machine conversation by the visual system, the auditory system and the body system at the same time [2]. The spatial metaphor, based on people's physical behavior, shows great tendency and regularity in people's expressions [3], which means spoken language and gestures are the main ways of people's expressions.

The main purpose of this paper is to explore the correlation between the expression of human gestures and spoken language, and the results of the study are used to reduce the conflict between users and designers. This paper will discuss the usage of gesture from the perspective of 3D space, explore the users' cognition from the aspect of space metaphor, and find the relationship between schema of oral report and gesture expression. By trying to construct a system to establish spatial perception model through a combination of schema and gesture paths, an auxiliary communication platform between users and designers will be built.

2 Related Research

At present, a great number of research probe into this issue and different tools have been adopted to explore the communication between designers and users. In the field of UCD research, Han et al. [4] studied and expounded the UCD of the product design area from the perspective of usability. Chang et al. [5] took cars as the case adopted the method of semantics for users' feedback form about the product demand, and focused on the expression pattern of the users, which have special meanings for understanding users' requirements. Chen et al. [6] explored the features of different parts of the car's modeling and their roles in users' perception from the perspective of the modeling's characteristics. At the same time, the designers need to integrate information and create products. Therefore, to improve the credibility of product design innovation, it is very important to understand the designers' thinking characteristics and cognitive styles. The latest progress of Intel company in 2013 has established a special perception computing department, which tries to combine non-contact gesture control with face recognition, voice command, eye tracking and other new technologies together to form a computer unified solution for the natural input. Based on these theories, this paper tries to explore the relationship between the non-contact gesture and the voice command.

3 Experiment

3.1 Purpose

Two experiments are designed in this part for observing the internal relationship between gesture and oral report when users express their requirements and design spatial paths according to oral reports, as to improve the collaborative awareness between users and designers.

The purpose of the first experiment is to find relationships between gestures and oral reports of users when they express their requirements, and to compare gestures in different contexts with the same image schema word. The aim of the second experiment is to design spatial gesture paths according to image schema of spoken language and to verify the consistency between the gesture path and the corresponding image statement.

3.2 Method

In this paper, image schema is used to abstract diagrams from oral reports and taking as foundations to design gesture. Image schema, as an abstract structure, comes from the activities between the human body and the outer space, which is the "skeleton" of many activities with common features, is abstract and invisible in the minds, is constantly reproduced from human activities, can be used to organize the human experience and connect the related experience together, is produced from the specific experience of human beings, because humans can map it to the abstract concept, therefore it can be used to recognize people's abstract concept [1]. Besides, image schema can be generally defined as dynamic analog representations of spatial relations and movements in space [7] and image schema can be represented by schema diagram according to active experience [1]. Therefore, humans' activities could be abstracted into image diagrams by image schema. It was Lakoff and Johnson who put forward the concept of image schema theory first. Johnson explored the experiential basis of image schema in the whole book. In this paper, the 28 schema words which were put forward by Johnson are used for abstract participants' gestures [8].

20 students were invited in the first experiment as the users, expressing the requirements of A Cup. The oral report and the corresponding gestures were picked out and be abstracted with image schema words proposed by Johnson.

The second experiment is based on the first experiment. The most common 5 image schema words (PATH, CONTAINER, PARTWHOLE, CONTACT, BALANCE) in Experiment 1 were selected. Each word was given an image diagram and corresponding statement according to the theory [1], the theory introduces that most of human activities could be described by image diagrams according to the cognitive linguistics and the life experience. 15 hand gesture designers were invited to design gestures according to the given words and the corresponding statements. 5 gesture paths were determined by comparing the gesture paths and image diagrams. In order to verify the consistency between the gesture paths and the meaning of the image statement, 95 users were invited to make scores on a 5-point Likert.

3.3 Experiment

Users' Gestures and Image Schema of Oral Reports. In the first experiment, 20 students (11 Females, 9 Males) whose average age is 22 years old (N = 20, SD = 2.12), and are from different majors participated in the first experiment. These 20 subjects were divided into 5 groups. The topic was glass design, subjects decided cup appearance and determined in advance and described their needs within 15 min.

During the experiment, each group with the camera to record the user's gestures and description. After the experiment, the users' gestures and oral reports were extracted and sorted according to the image schema.

Johnson put forward more than 20 image schema words: CONTAINER; BALANCE; COMPULSION; BLOCKAGE; COUNTERFORCE; RESTRAINT; REMOVAL; ENABLEMENT; ATTRACTION; MASS-COUNT; PATH; LINK; CENTER-PERIPHERY; CYCLE; NEAR-FAR; SCALE; PART-WHOLE; MERGING; SPLITYING; FULL—EMPTY: MATCHING: SUPERIMPOSITION; ITERATION; CONTACT: PROCESS; SURFACE; OBJECT; and COLLECTION. These schema words are the most common schema of humans' activities. Therefore, in the experiment, the user's description of their external activities are marked and matched with part of the graphic words which were summarized by Johnson. These words were selected according to the cognitive linguistic rules and semantics. At the same time, if the users use gestures to assist their expressions, gestures are recorded, and be divided into different types of image schema.

Correlation Between Spatial Gesture Paths and Image Schema. From the first experiment, several image schema are abstracted from users' gestures. In order to further prove the relationship between these image schema and spatial gesture path in the experiment 1, the second experiment chose the five image schema with the highest frequency from the first experiment as the experimental basis. According to these five image schema, there are 5 schema statements and the 5 corresponding schema diagram, 2 schema were chosen from the existing image schema theory [1], and other 3 schema were created according to the conceptual metaphor and image schema approach [9] (Table 1).

15 designers with two years gesture design experience were invited into the second experiment, and subjects were asked to read 5 schema statements and designed the corresponding gestures at the same time. In order to facilitate the path extraction, participants were invited with black clothes, and designers created gestures before the black background. Three cameras recorded the whole experiment process from three different directions. 5 gesture paths were determined by comparing the gesture paths and image diagrams from 15 gesture designers. In order to verify the consistency between the gesture path and the meaning of the image statement, 95 users were invited to experience and make scores on a 5-point Likert. 15 gesture designers were numbered U1–U15. Gestures of each image schema were numbered Ux (1)–Ux (5).

3.4 Analysis

Oral reports of the first experiment are divided into four categories: description of spatial form, style orientation, functional requirements, and case enumeration. Sentences of each category have different image schema according to the language context. The following is part of the extraction from the oral reports and corresponding gestures (Tables 2, 3, 4 and 5):

Table 1. Schema statements and schema diagrams.

Image Schema	Statement	Diagram
PATH	Look at a train, from the first carriage to the last	
CONTAINER	I pour out the milk from the bottle	
PART-WHOLE	I drive the sheep back to the flock	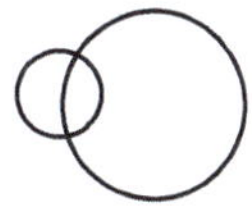
CONTACT	The Internet connects us closely together	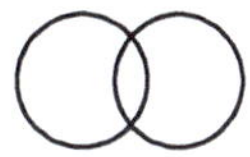
BALANCE	Eating meat and vegetables can balance nutrition	

Table 2. Description of spatial form.

Oral Report	Image Schema	Gesture
"I like a small cup, thick cup is hard to carry"	CONTACT	
"Cups with round outline could carry more water."	CONTAINER	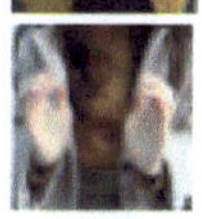
"I love cups which can set up a complete shape."	PART-WHOLE	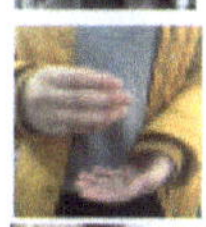
"I like cups with axial symmetry outline."	BALANCE	

Table 3. Style orientation.

Oral Report	Image Schema	Gesture
"I like cups with simple style."	CONTAINER	
"I like to buy cups with similar style for my family, it looks enjoyable."	PATH	
"I prefer the cup with some small texture, it seems full of design sense."	CONTAINER	
"I love the wood and purple collocation, looks very harmonious."	BALANCE	

Table 4. Functional requirements.

Oral Report	Image Schema	Gesture
"It's very convenient to carry a cup which with a rope to everywhere."	PATH	
"I prefer use thermos cup, cause I usually drink warm water."	CONTAINER	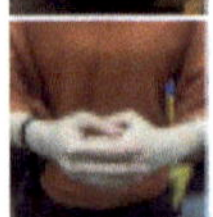
"I prefer the cup with a small cup so as to avoid being scalded."	PART-WHOLE	

These image schema words above were all derived from the part of 28 schema vocabularies by Johnson. And image schema vocabularies in tables were abstracted from users' oral reports according to the schema and meaning of activity descriptions. In order to compare different gestures with the same image schema in different context, gestures with the same image schema were picked out. The following Table 6 is about the image schema and the corresponding gestures:

At this stage, gestures with the same image schema were coordinated. Paths were extracted according to the paths of their fingertips and the movement directions were showed by arrows. Spatial gestures were recorded by three cameras from three different directions which standing for X, Y, Z axis respectively to build spatial paths models according the method from Tan [11].

The following Table 7 is about gestures which designed by subjects according to the given schema statements in the second experiment:

Table 5. Case enumeration.

Oral Report	Image Schema	Gesture
"I like MUJI cups, I often go to the MUJI store."	ATTRACTION	
"My girlfriend gave me a THERMOS cup, so I am very interested in this brand."	CONTACT	
"I had a cute mug, but It was broken and I bought the same mug again."	ITERATION	
"I like lovely cups, so I look for lovely cups everywhere.	COLLECTION	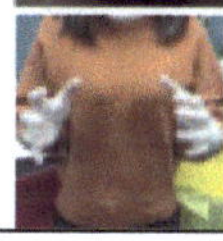

Table 6. Gestures and image schema.

Image Schema	PATH	CONTAINER	PART-WHOLE	CONTACT	BALANCE
Gestures	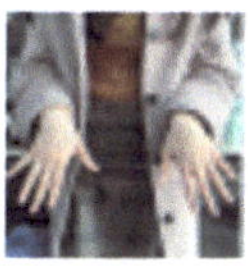				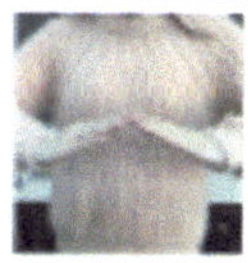

Spatial Paths					

Table 7. Schema statements and gestures.

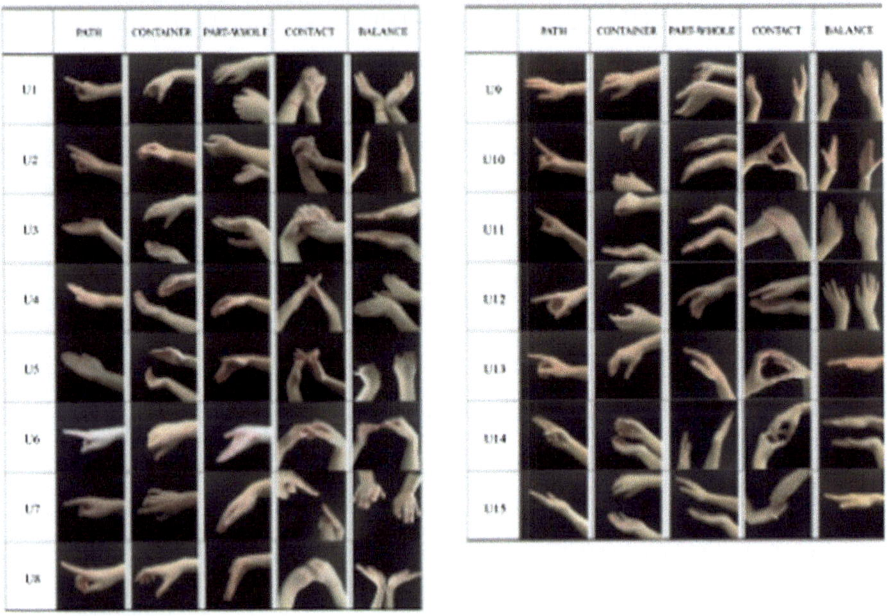

After collected users' gestures, spatial paths of gestures were abstracted and the paths on X axis were chosen to compare with image schema diagrams. The following Table 8 is about the image schema diagrams and corresponding gesture paths:

Table 8. Image schema diagrams and corresponding gesture paths

PATH	CONTAINER	PART-WHOLE	CONTACT	BALANCE

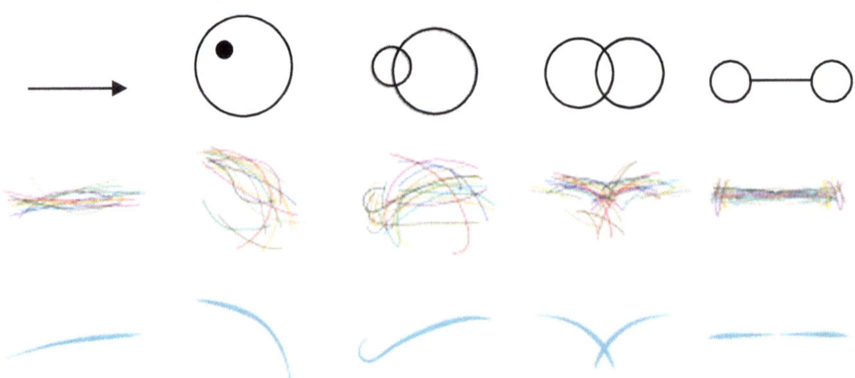

By comparing the schema diagrams and spatial paths of gestures, Table 8 shows that the image schema diagram has consistency with spatial paths of gestures. In order to verify that the gesture spatial path is consistent with the user's perception, participants consisted of 42 males and 53 females who ranged from 20 to 28 years old (N = 95 SD = 2.48) were invited to test those 5 spatial gesture paths by repeating schema statements with the given path. Participants need to score the complete experiment process on a 5-point Likert scale ranging from 1 (very poorly) to 5 (very well) to evaluate the perceived identify of spatial paths.

4 Results

4.1 Result

It shows that females prefer to use sign language to express their requirements and the average time for them to use gestures is 6.3 min. There were 14 participants using gestures, which means more than half users rely on sign language. The first experiment results show that the users' gestures have the relevance and consistency for the same image schema. The approximate spatial path is similar, although the postures of the hand are slightly different (Table 9).

Table 9. Gesture users and cumulative using time

Females	U2	U3	U4	U5	U7
	3'11	10'21	6'12	6'51	5'33
	U10	U14	U17	U19	–
	9'06	3'08	7'26	9'42	–
Males	U1	U9	U11	U12	U20
	4'05	9'29	9'08	4'24	4'32

The above line chats are about the scores of five spatial paths. Scores are from 1 (very poorly) to 5 (very well), and the score higher than 3 stands for medium standard. The experimental data shows that the line of the "path" is the flattest, the scores are mainly concentrate in 4 and 5, and the average score is 4.41. The data shows that the users' perceptive conflict of this spatial path is minimal. The average score of all spatial paths is 4.084, which can prove that the users have small perceptive conflict of the spatial gesture paths and the corresponding schema statements (Figs. 1, 2, 3, 4, 5 and 6).

The data shows that the number of objects concentrating in 4 and 5 of each spatial paths are 84, 54, 68, 73, and 86 respectively. The spatial path of the "BALANCE" has the most objects with good experience, but the spatial path of the "PATH" has the highest average score, which means the perceptive conflict of "PATH" is the lowest among these five spatial paths but the quality of the experience of the "BALANCE" is the best. After the research, according to several designers whose score concentrating in 4–5 of the "BALANCE", the main reason is the image "BALANCE" showing a torsion balance in their mind. Hence the image schema of the users' spoken language

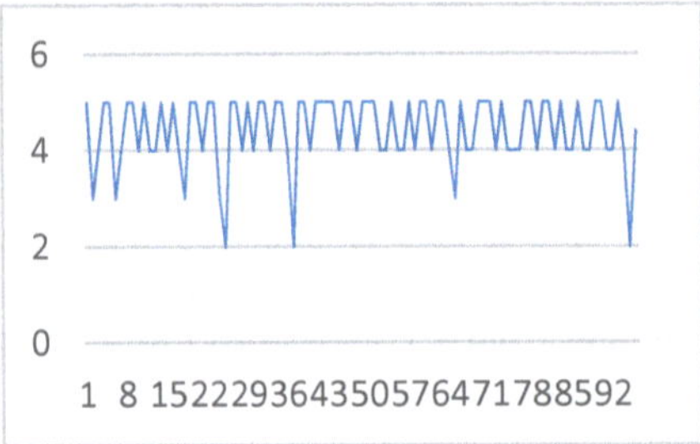

Fig. 1. Scores of spatial paths "path"

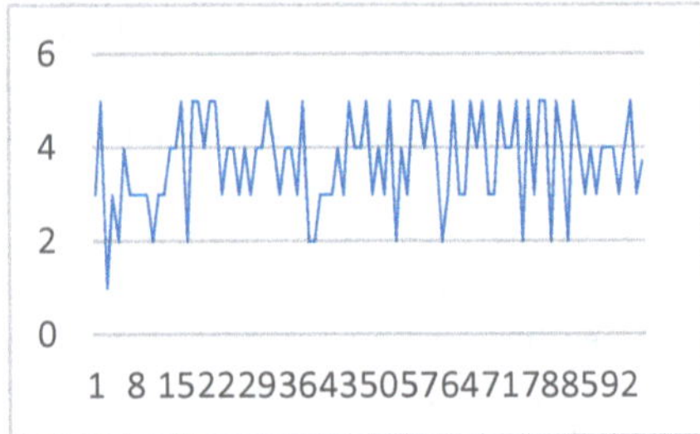

Fig. 2. Scores of spatial paths "container"

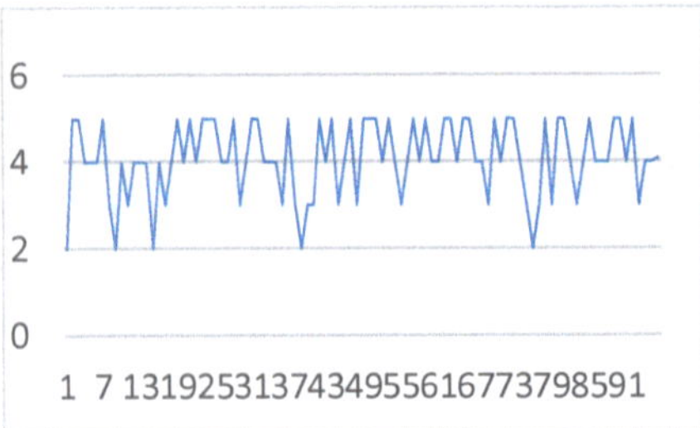

Fig. 3. Scores of spatial paths "part-whole"

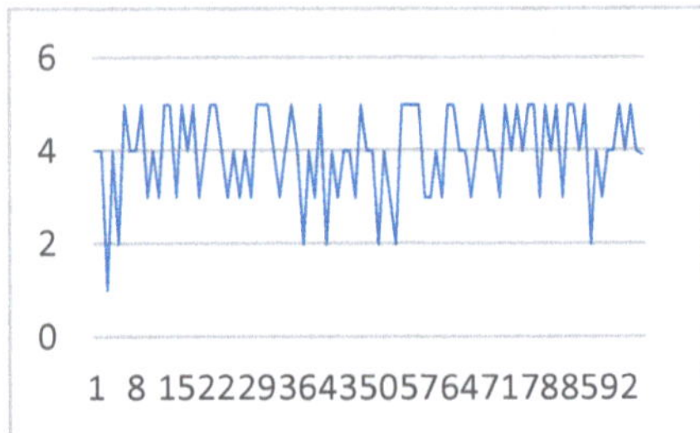

Fig. 4. Scores of spatial paths "contact"

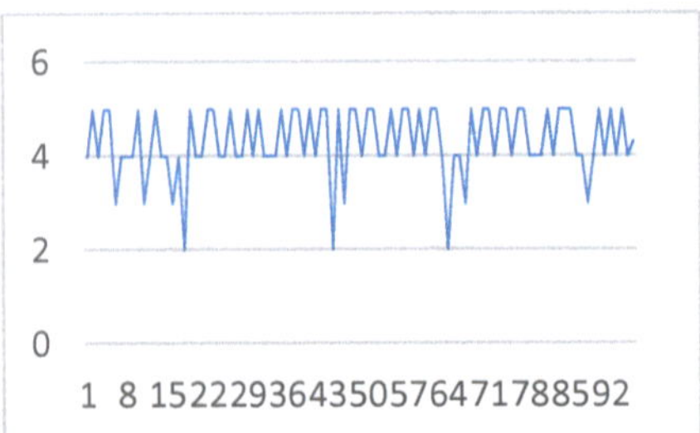

Fig. 5. Scores of spatial paths "balance"

Spatial path	Average score
PATH	4.41
CONTAINER	3.70
PART-WHOLE	3.92
CONTACT	4.08
BALANCE	4.31

Fig. 6. Average scores of spatial paths

and the spatial gesture paths have consistency which could reduce the perception conflict between users and designers.

It shows in these two experiments that users rely on gestures and oral expressions during the communication. Being from human's experience with the outside world, image schema is an important concept in cognitive linguistics, and can be expressed by gestures and spoken language. The research has certified the consistency of the oral report and spatial gestures by the image schema, and users will have high level perception identification with spatial paths abstracted from oral reports. Therefore, if the users' gesture paths follow spatial lines which are similar to image schema diagrams, designers will understand the users easily.

4.2 Discussion

There are still many deficiencies in this paper: firstly, the first experiment only selected parts of the image schema, so the conclusion cannot be completely suitable for all image schema; secondly, the corresponding path diagrams were created by only parts of designers, but there should be many other participants to be included; thirdly, participants are Chinese, and the image schema were carried out in the context of Chinese. Therefore, the conclusion may only be applicable to Chinese language. Furthermore, there are numerous work needed to do on the expression of image schema with spatial sign language.

- Invite foreigners to participate in the experiment to compare the similarities and differences of the image schema between Chinese and English.
- Combine the Unity 3D and Leap Motion to generate a speech gesture recognition system which can build models by capturing users' gestures.
- Involve more image schema, abstract more spatial gesture paths, and collect spatial paths together as a set, so as to reduce the cost of awareness conflict.

Acknowledgments. We would like to express our gratitude to Li Fuyin who helped us during the research. The research was supported by National Natural Science Foundation of China (61402159/60903090), National Key Technologies R&D Program of China (2015BAH22F01/2015BAH22F02), Hunan Provincial Social Science Foundation of China (2010YBA054), the State Key Laboratory of Advanced Design and Manufacturing for Vehicle Body Funded Projects.

References

1. Fuyin, L.: On image schema theory. J. Sichuan Int. Stud. Univ. **23**(1), 80–85 (2007)
2. Wang, G.-J., Wang, Z.-L., Yang, G.-L., et al.: Survey of artificial emotion. Appl. Comput. **23**(11), 7–11 (2006)
3. Lakoff, G., Johnson, M.: Metaphors We Live By. The University of Chicago Press, Chicago (1980)
4. Han, S.H., Yun, M.H., et al.: Evaluation of product usability: development and validation of usability dimensions and design elements based on empirical models. Int. J. Industr. Ergon. **26**, 477–488 (2000)
5. Chang, H.-C., Lai, H.-H., Chang, Y.-M.: Expression modes used by consumers in conveying desire for product form: a case study of a car. Int. J. Industr. Ergon. **36**, 3–10 (2006)
6. Chen, L.L., Kang, H.-C., Hung, W.-K.: Effects of design features on automobile styling perceptions. In: International Association of Societies of Design Research, Hongkong, China, November 2007
7. Gibbs, R.W., Colston, H.L.: The cognitive psychological reality of image schemas and their transformations. Cogn. Linguist. **6**, 347–378 (1995)
8. Johnson, M.: The Body in the Mind: The Bodily Basis of Meaning, Imagination, and Reason. The University of Chicago Press, Chicago (1987)

9. Li, T.F.: The acquisition of metaphorical expressions, idioms, and proverbs by Chinese learners of English: a conceptual metaphor and image schema-based approach. Ann Arbor, Michigan: UMI. ProQuest Information and Learning Company, p. 366 (2003)
10. Nathan, C., Moultrie, J., Clarkson, P.J.: Seeing things: consumer response to the visual domain in product design. Des. Study, **25** (6), 547–577 (2004)
11. Tan, H., Li, W., Tan, Z.Y.: 3D gesture interaction product design of vehicle information system. Packag. Eng. **36**(18), 45–53 (2015)
12. Tan, Z.Y., Dai, W., Tan, H.: Three-dimensional gesture design based on imagery. Packag. Eng. **37**(4), 61–65 (2016)

Author Index

A.G. Ho (ed.), *Advances in Communication of Design*, Advances in Intelligent Systems and Computing 609, DOI 10.1007/978-3-319-60477-0

Author Index

MIX
Papier aus verantwortungsvollen Quellen
Paper from responsible sources
FSC® C105338

If you have any concerns about our products,
you can contact us on
ProductSafety@springernature.com

In case Publisher is established outside the EU,
the EU authorized representative is:
Springer Nature Customer Service Center GmbH
Europaplatz 3, 69115 Heidelberg, Germany

Printed by Libri Plureos GmbH
in Hamburg, Germany